THE ROLE OF PLANTATION CROPS IN AGRICULTURAL DEVELOPMENT

The Role of Plantation Crops in Agricultural Development

S. Giriappa

2022

Daya Publishing House®

A Division of

Astral International Pvt. Ltd.

New Delhi – 110 002

SOMU GIRIAPPA (b. 1944 –)

First Published, 1989

Reprinted, 2022

ISBN: 978-81-7035-057-6 (HB)

Published by : **Daya Publishing House**®
A Division of
Astral International Pvt. Ltd.
– ISO 9001:2015 Certified Company –
4736/23, Ansari Road, Darya Ganj
New Delhi-110 002
Ph. 011-43549197, 23278134
E-mail: info@astralint.com
Website: www.astralint.com

Digitally Printed at : **Replika Press Pvt. Ltd.**

Preface

Plantation crops have been playing a vital role in augmenting agricultural efficiency and value. Major plantations not only yield revenue to the state exchequer but also bring forth foreign exchange. Though coconut and arecanut are classified as palms, an attempt is made in this study to include them and study some other plantation crops like cashewnut to the development of agricultural and the rural sector in the particular region. The study is divided into two parts—Part I comprising coconut-arecanut and Part II cashewnut.

This study has been completed under the aegis of the SBI Chair in Rural Development, Mangalore University. The Cashewnut study was accomplished first and then the coconut-arecanut study. The Institutional Finance and Statistics Department, Government of Karnataka deserve special thanks for financial assistance for the Cashew field study. In the conduct of the field study, M/s Sripathi, K.P., Gopal, N. Ashok, P. and M. Seetharam helped in a great way. Ms. M. Muktha Shet and Sri M. Seetharam ara to be thanked for their assistance in processing and drafting and typing of the manuscript.

S. GIRIAPPA

Contents

1

Plantation Crops

Itnroduction

Plantation crops have been traditional export earners but their importance declined when industrial and engineering goods become major export products. Interest in them was revived through new technology and marketing methods which were made available after the wake of revolution. The concentration of plantation crops is found mainly in the third world countries and hence lack of research and development by advanced countries (unlike in the case of cereals) might have been the major reason for the plantation sector to have remained traditional inspite of being industrialised. To maintain the tempo of interest in them increased product consumption greater export potentiality and attempts on development of the particular crops have been strengthened.

Plantation crop development is another way of containing the environmental deterioration caused by the modern agro-systems. Normally, area under plantations does not constitute agricultural impoverishment, though it could replace some forest cover. Plantation and forestry could go hand in hand and toward this and could develop what is called a forestry-plantation system. Ecologically, the plantation sector is better endowed in stability and ecological sustainability. Plantation development if it is in wasteland, could propogate mixed crop species along with soil conservation and linkages with other sectors thus improving the role of plantation crops.

The National Commission on Agriculture had forecasted a decline in food crop area and increase in commercial crop area by 2000 A.D. In the matter of plantation crops, the area under cultivation was projected to increase from 2.38 mil ha in 1975 to 2.84 mil ha in 2000 A.D. Table 1.1 presents the area and target of some plantation crops. The area under more organised plantation crops like tea, coffee, rubber and cocoa has to increase, while that of less organised crops like arecanut and coconut has to remain more or less constant. Similar is the case with cardom and pepper. Introduction of oil palm and cashew development are other aspects of the growth strategy. Setting up of individual commodity Boards and instituting a viable export performance especially for the organised plantation crops would also help in their regeneration.

The main plantation crops have been identified to include not only the organised crops but also energy plantation, spices and condiments, oilseeds, forest plantation and other tree plantation crops. Development of indigeneous species along with exotic hybrids in line with local requirements places greater emphasis on social forestry and oilseeds plantation. The new genetic development in crops with tissue culture and other modern technology aspects have tried to spread new species to serve different purposes like energy, biomass, ecology, productivity and growth.

Productivity of Plantation crops

The major plantation crops tea, coffee, rubber and cardamom share less than 0.5 per cent of total cropped area, but account for over 10.0 per cent of value added in agriculture. Tea and rubber have performed satisfactorily during the last 2 to 3 decades. Between 1950 and 1980, the share of tea in internal consumption increased from 25.5, to 63.0 per cent, while that of coffee fell from 52.0 to 35.4 per cent. The incidence of frost in Brazil in the mid-seventies and the resulting loss in coffee production and natural calamities in Columbia thereafter were boons to Indian coffee export in certain years. Though the proportion of coffee in the export market has increased, Indian share in world market is very negligible. On the other hand, India is a major producer and exporter of tea, though the share is declining due to competition from south-east Asia and Kenya. During the mid-seventies the government policy of the constraining export had resulted in sharp increase in internal consumption since then. However, the average per capita consumption of tea and coffee (400 and 220 gms per year) is less than a tenth of the world average.

The productivity of tea has increased from 849 kgs/ha in 1950-51 to 1477 kgs/ha in 1980-81, one of the highest in the world. During 1950-51, 72.2 per cent of the total tea production was exported, whereas the share declined to 38.9 per cent in 1980-81. The area increased from 0.32 to 0.37 mil/ha during this period. Coffee has had good performance in export, the share going from 14.4 to 74.2 per cent of total production in this period. Its area increased from 0.1 to 0.2 mil/ha.

Rubber has registered a phenomenal growth (over three times on the base) and its area increased to 0.24 mil ha in 1980-81. The country was able to reduce rubber imports from 21.0 to 5.3 per cent of total internal consumption in 1980-81. The drought after 1980-81 had affected the productivity of all the crops including rubber.

In the case of cardamom, black pepper and cashewnut, the yield rate is declining rapidly impairing the export performance. In coconut, there has been a slight decline in productivity at the national level, whereas in arecanut the yield level has been increasing. Table 1.2 describes the productivity of major food and commercial crops during 1971-72 to 1985-86 at the national and regional levels. This is to bring out the comparison between green revolution crops and plantations which are experiencing technologisation recently. Rice and ragi have shown increase in productivity during the period both at the national level and in Karnataka. Commercial crops like groundnut and tobacco follow an increasing trend at the national level only. Tea, coffee, rubber and arecanut have registered increased decline whereas Karnataka has registered an increase. In the case of cardamom and balck pepper, the productivity shows better performance but lower when compared to 1950-51 level. Cashewnut productivity has been declining rapidly when compared to other plantation crops. However recent developments in the crop have enabled to realise an increasing productivity. It may be seen that the problem with some plantation crops is similar to that experienced by pulses and oilseeds to a certain extent.

Crops like rice, ragi, sugarcane, tobacco and banana have registered a growth of about 1.5 per cent per annum, whereas there has been a decline in groundnut yield to that extent. Plantation crops like coffee and cardamom have registered a growth rate of over 3.0 per cent per annum, though in the case of tea, rubber and pepper, the growth pattern has been similar to that of non-plantation crops.

In Karnataka, there has been a substantial increase in yield rates of ragi and tobacco and in groundnut, sugarcane and banana negative growth has been registered. Among plantations, coffee, rubber and arecanut have fared very well but negative performance in cardamom (3.0 per cent) and pepper (about 45.0 per cent) was visible. On balance, it seems that plantation crops have also fared satisfactorily similar to the green revolution and commercial crops. Their performance was quite satisfactory when compared to that of millets, pulses and oilseeds both at the national and state levels.

Scope and Method

This study attempts to analyse factors in the growth of coconut, arecanut and cashewnut with special reference to Dakshina Kannada and Karnataka. Issues like area and productivity, changes over the recent years, cultivation practices and evaluation of new programmes are being considered. In Karnataka, coconut plays an important role not only in effecting a better ecological agro-system, but also in meeting fuel and oil requirements and many other socio-economic uses. Area under coconut has been increasing in recent times and has assumed a great commercial value. Arecanut is a important crop in the irrigated and coastal districts. The World Bank-aided cashew project attempts to increase cashewnut production toward greater export performance.

Besides these, plantation crops like tea, pepper, cardamom, rubber and cocoa are also considered briefly. In tea, the main analysis is with energy consumption. Altogether, the study may be divided into three parts. The first part dealing with coconut and arecanut, is based on a field study wherein comparitive information was gathered. In doing so, particulars of other crops especially paddy were also covered. And hence this part would depict the paddy and non-paddy farms cultivating coconut and arecanut in the district.

In the second part, minor plantation crops and energy consumption in the Nilgiri tea industry are considered and the analysis is mainly based on secondary information in these cases. The third part dealing with cashewnut development in the district attempts to study the performance of beneficiaries of the World Bank aided project and the non-beneficiaries so as to bring about the economic evaluation of the former. In doing so, a comparison of other crops grown along with cashewnut is bound to be practised. Here, crops like coconut, arecanut and paddy are dealt with briefly in so far as they depict the different comparitive evaluations of the project and the non-project cashew cultivators.

Plantation crop development assumes great importance in the sphere of export performance. Recent World Bank studies point out that economic growth would not be speeded up unless due attention toward exportable surplus is paid. When decline in traditional exports is experienced (and severe competition for modern goods and services), revival of cheep traditional industries where the country has specific raw material capability might be a right step in the direction. At the same time satisfaction of internal demand is another factor that has to be given due consideration. While some plantation crops (palms) try to fulfil internal demand (actually in coconut there has been an import spree due to its poor performance), others try to improve their export performance (instances of coffee and tea). Some other crops though having great export potentiality are plagued with low productivity and unstable prices (instances of pepper and cardamom).

The study attempts to depict the national, international (wherever the case may be), regional and local level performance of the plantation crops in terms of area, production and yield based on secondary information. Wherever possible primary information has been provided to substantiate the macro-level performance or to elucidate the particular local specificity as to the nature of prospects the particular crop has. It may be pointed out that some important plantation crops such as coffee have been kept out of the purview of this study.

TABLE 1.1: Targets of area for plantation crops-2000 AD

Crops	*Area (mil ha)*	
	1974-75	*2000-01*
Tea	0.35	0.45
Coffee	0.14	0.20
Rubber	0.20	0.38
Cardamom	0.076	0.08
Cocoa	0.001	0.015
Arecanut	0.17	0.17
Coconut	1.05	1.05
Cashewnut	0.22	0.35
Oil palm	—	0.025
Pepper	0.12	0.12
Total	2.327	2.840

Source: Anon. *Report of the National Commission on Agriculture, 1976 Part IV,* Crop Production, Sericulture and Apiculture, Appendix Table 24.4.

TABLE 1.2: **Productivity of major crops (kg/ha)**

Crops	*1971-72*		*1975-76*		*1980-81*		*1985-86*	
	1	*2*	*1*	*2*	*1*	*2*	*1*	*2*
Rice	1141	1872	1235	1892	1338	2037	1402	1943
Ragi	911	895	1064	1166	1053	1275	1286	1414
Groundnut	823	756	935	655	727	597	858	953
Sugarcane	47511	82171	50993	75572	56844	81714	57600	79058
Banana	15016	8374	14732	5606	16685	5417	15659	5521
Tobacco	914	419	950	603	1064	624	1089	651
Tea	1271	1766	1405	1950	1477	1884	1777	1914
Coffee	498	614	490	600	702	909	787	983
Rubber	678	412	772	536	771	632	821	682
Coconut (nuts)	5626	4838	5449	5073	5249	5176	5093	5190
Arecanut	843	1229	901	1418	1037	1459	994	1462
Cardamom	41	61	52	65	55	59	54	57
Cashewnut	837	846	325	385	497	309	520	308
Pepper	221	476	229	265	248	259	250	242

1=India 2=Karnataka

Source: Anon. *Estimates of Area and Production of Principal Crops in India*, Directorate of Economics and Statistics, Ministry of Agriculture, New Delhi, various issues.

PART I

2

Introduction to Coconut-Arecanut Cultivation

Table 2.1 presents land use and crop patterns in the region. In Dakshina Kannada, 27.1 per cent of total geographical area is under forest, while the proportion is only 16.0 per cent in Karnataka. About 54.0 per cent of area constitutes net area sown in the state while the proportion is only 26.0 per cent in Dakshina Kannada. During 1986, the actual rainfall in both the state and district was lower than the normal by about 11.0 and 14.0 per cent respectively. The proportion of net irrigated area to net area sown is higher in Dakshina Kannada (39.6 per cent) but the proportion of gross irrigated area is only 23.5 per cent. Karnataka has lower proportion of 16.4 per cent net and 18.1 per cent gross irrigated area. The proportion under rice, coconut, arecanut and cashewnut are quite significant in Dakshina Kannada at 52.2, 6.1, 6.0 and 8.8 per cent respectively during 1985-86. In Karnataka, the proportion of area under these crops is not significant (especially in the case of crops other than rice). In Dakshina Kannada, over 73.0 per cent of gross cropped area is under these four crops, while at the state level, they share only about 13.0 per cent. Because of heavy rainfall, rice is cultivated mainly in the kharif season though rabi area is also significant (28.6 per cent as against 66.7 per cent Kharif). The proportion of marginal and small farms is over 60.0 per cent of total landholdings, much higher than the state level. The area per land holding is about 2 ha in the state and 1.8 ha in the district. In the district, except Sullia all other taluks cover substantial area under rice. In coconut, the concentrated taluks are Udupi, Karkal, Bantwal, Belthangady and other taluks. Bantwal, Sullia, Puttur and Belthangady report higher arecanut area. In cashewnut,

Karkal, Udupi, Kundapur and also Bantwal, Puttur and other taluks report substantial area coverage.

The Sample

The sample area for evaluating the performance of coconut, and arecanut has been selected in the Puttur taluk, which has concentration of arecanut and also coconut to some extent. A comparison of paddy, coconut and arecanut cultivation in respect of the sample farmers would indicate the nature of resource allocation and environmental management. Arecanut gardens were originally paddy fields and the rising commercial value of arecanut (also infused by the successful functioning of the Central Arecanut and Cocoa Marketing and Processing Cooperative Society) in the recent past had induced to the farmers to switch over to arecanut cultivation. Because of this, two types of farmers were encountered those who grow paddy along with arecanut or coconut and the those who grow only arecanut and coconut. The arecanut gardens are usually mixed with banana, cocoa, fruit trees inside and coconut in the borders. In some gardens coconut has been mixed with arecanut but in cases of young coconut palms certain growth problems were encountered. The recent practice has been discouragement of arecanut-coconut mix and cultivation of coconut on the borders only. After a certain years of growth areca seedlings will be planted in between the older palms.

In paddy fields also, coconut is planted on the bunds and borders only. Depending upon water availability, a second crop of paddy or pulses would be cultivated in the paddy field. During this season, certain vegetables will also be grown. In the summer the land is left as fallow or if water is available a vegetable crop will be raised. In few cases a second paddy crop will be cultivated.

Table 2.2 describes the distribution of the sample units according to different land holding groups. The number of farmers cultivating paddy were 35, while those not cultivating numbered 20. The land holding size includes wet land, garden land and other land (cashewnut, pepper, pulses etc). In the sample area (Ariyappu and adjoining areas of Puttur taluk), the group classification has been less than 1 ha, 1-2 ha, 2-3 ha and 3-4 ha of combined wet, rainfed and garden lands. The groupings have been separately attempted for paddy cultivating and non-cultivating farmers. In the case of paddy farmers the crops cultivated were arecanut, coconut, banana, pulses like pigeon pea, blackgram, greengram, rubber, pepper, cocoa, etc, besides paddy. In the second grouping the farmers grew crops other than paddy.

The average age of the farmers in the first category was 53.3 years while it was lower at 43.8 years in the second category. The average family size in group I was 7.5 and in the II 6.3. The family size was above the average in the I and III size groups of first category and II and IV size groups of second category. As to employment particulars in and outside agriculture the average male and female workers were lower in number, when compared to the smaller size groups. On average, 1.2 and 1.6 male workers were engaged in agriculture and non-agricultural activities in the first category. The III group in the second category and III group in the first category reported a larger working population when compared to the category averages. The last group had a lower proportion of workers. Participation of women in non-agricultural activities was insignificant in both the categories. In the second category, participation by men in non-agricultural activities was not significant as in the first. For the entire sample the average age of the household head was 49.2 years. The average family size was 7 and about 47.0 per cent of them were workers (60.0 per cent in agriculture and 40.0 per cent in non-agriculture activities).

Table 2.3 presents the cultivated area and cropping pattern particulars of the different size groups in the two categories. In the paddy category, the average land-holding was 2.50 ha per household. It was 0.96 ha in the I group, 1.67 ha in the II, 2.8 ha in the III and 3.94 in the last group. Of the total land area about 30.0 per cent was irrigated, the rest being garden land mostly irrigated. The proportion of paddy land was lower in the first category (31.6 per cent) but in the II and III groups of that category, the proportion of the paddy area was higher at 36.6 and 34.5 per cent respectively. The proportion of arecanut in the garden land was substantial, 64.2 per cent of total garden land for the entire category, 58.6 per cent for the I group, 62.2 per cent for II group, 61.3 per cent for the III and 69.0 per cent for the IV group.

In the non-paddy cultivating category, the average land size was 2.45 ha; it was 0.98 in the I, 1.95 ha in the II, 2.68 ha in the III and 3.65 ha in the IV groups. The proportion under garden crops was over 60.0 per cent, much higher than in the paddy cultivating category. The proportion under arecanut and also the net cultivated area were higher in this category.

For the entire sample, the average land-holding area was 2.48 ha. The proportion under different crops was other crops 29.5 per cent, arecanut 32.7 per cent, paddy 20.1 per cent and coconut 17.7 per cent. The proportion under arecanut was higher than the sample average under I and II groups (over 35.0 per cent), nearing average in the IV group (33.2 per cent) and lower than the average in the III group (29.3 per cent). The average paddy area was 20.0 per cent for the entire sample and higher for the III group (over 32.0 per cent). The proportion under coconut cultivation was around 20.0 per cent in I, 19.0 per cent in II, 18.0 per cent in III and 16.0 per cent in IV groups. The area under other crops was less than the average in I and II groups (23.7 and 24.0 per cent) and higher in the III group and very low in the IV group at 16.2 per cent.

It may be seen that under both the categories the average landholding in group I was more or less similar. The landholding area in groups II and IV of the second category and group III of the first category was slightly higher than the counterparts in other category. The higher proportion of garden land in the second category is worthwhile to note especially with reference to problems of paddy cultivation in the region. Heavy rainfall and related factors have favoured rainfed paddy cultivation but over years the proportion of paddy area to total cropped area has been steadily declining. For example, paddy accounted for about 73.0 per cent of total cropped area in 1955-56 but the proportion declined to about 52 0 per cent after 30 years. In this region, the increase in high yield rice strains was also very marginal-from 49.2 per cent in 1978-79 to 51.1 per cent in 1985-86 indicating the risky nature of paddy cultivation. It could be seen that the proportion of area under coconut, arecanut and cashewnut has significantly increased during the above period-from 4.5, 2.5 and less than 5.0 per cent to 6.2, 6.1 and 12.5 per cent respectively.

TABLE 2.1: **Land use and crop patterns**

(area in ha)

Items	*Dakshina Kannada*	*Karnataka*
Total geographical area	8,33,595	1,90,49,836
Forest area	2,25,755	30,57,030
Net sown area	2,17,057	1,01,72,192
Gross area sown	29,7,919	111,46,388
Net irrigated area	86,031	16,74,996
Gross irrigated area	99,697	20,12,291
Area under rice	1,55,775	10,96,021
Area under Hyvs	80,320	25,78,293
Area under coconut	18,014	2,05,619
Area under Arecanut	17,941	58,640
Area under Cashewnut	26,194	63,000
Normal annual rainfall (mm)	3932	1355
Actual rainfall (mm)	3368	1209

TABLE 2.2: **General Particulars of the sample units**

Size group	*No*	*Age of h/h head*	*Total family members*			*Family Workers*			
						Male		*Female*	
			Male	*Female*	*Child*	*Agrl*	*Non Agrl*	*Agrl*	*Non Agrl*
I Paddy:									
Less than 1 ha	6	65	2.5	3.0	3.0	1.5	1.0	2.5	—
1—2 ha	9	48	1.5	1.5	3.3	1.0	0.8	0.8	0.5
2—3 ha	12	54	3.0	2.7	3.3	1.5	2.8	1.3	—
3—4 ha	8	45	2.0	1.6	2.0	0.5	0.5	0.3	—
Total average	35	52.3	2.3	2.2	3.0	1.2	1.6	1.2	0.2
II Non-Paddy:									
Less than 1 ha	3	53	1.0	1.0	2.0	1.0	—	1.0	—
1—2 ha	6	43	1.7	1.6	3.7	1.0	0.3	0.3	—
2—3 ha	6	39	1.7	1.3	2.3	0.8	—	0.5	—
3—4 ha	5	45	2.2	2.0	3.8	1.0	0.3	0.5	0.3
Average	20	43.8	1.7	1.5	3.1	0.9	0.2	0.5	0.1

III Total:									
Less than 1 ha	9	61	2.0	2.3	2.7	1.3	0.7	2.0	—
1—2 ha	15	46	1.6	1.5	3,5	1.0	0.6	0.6	0.3
2—3 ha	18	49	2.6	2.2	3.0	1.3	1.9	1.0	—
3—4 ha	13	45	2.1	1.8	2.7	0.7	0.4	0.4	—
Total	55	49.2	2.1	1.9	3.0	1.1	1.1	0.9	0.2

TABLE 2.3: **Cropping pattern in the sample units**

Size group	*No*	*Cropped area (ha)*				*Cropping pattern (in ha)*			
		Dry	*Wet*	*Garden*	*Total*	*Rice*	*Coconut*	*Arecanut*	*Others*
I Category									
Less than 1 ha	6	0.25	0.30	0.41	0.96	0.30	0.17	0.24	0.25
1—2 ha	9	0.32	0.61	0.74	1.67	0.61	0.28	0.46	0.32
2—3 ha	12	0.79	0.91	1.19	2.89	0.97	0.46	0.73	0.79
3—4 ha	8	1.12	1.11	1.71	3.44	1.11	0.53	1.18	1.12
Average	35	0.66	0.79	1.06	2.50	0.79	0.38	0.68	0.66
II Category									
Less than 1 ha	3	0.19	—	0.79	0.98	—	0.25	0.54	0.19
1—2 ha	6	0.58	—	1.37	1.95	—	0.43	0.94	0.58
2—3 ha	6	1.05	—	1.63	2.68	—	0.61	1.02	1.05
3—4 ha	5	1.46	—	2.19	3.65	—	0.75	1.44	1.46
Average	20	0.88	—	1.57	2.45	—	0.54	1.03	0,88

Average									
Less than 1 ha	9	0.23	0.20	0.54	0.97	0.20	0.20	0.34	0.23
1—2 ha	15	0.43	0.37	0.99	1.79	0.37	0.34	0.65	0.43
2—3 ha	18	0.88	0.61	1.34	2.83	0.61	0.51	0.83	0.88
3—4 ha	13	1.25	0.68	1.81	3.82	0.68	0.62	1.27	1.25
Average	55	0.73	0.50	1.25	2.48	0.50	0.44	0.81	0.73

3

Coconut

The coconut palm (*Cocos nucifera Linn*) is an old world crop long established in South East Asia and then spreading to other parts of the tropic. It is characterised by comparitively slow grow, thickening at the base of the trunk, and branched compact crown of feathering, glossy thick leaves at base and a branched infloresence inclosed in a sheath known as the apadix and absence of toproot and the continually growing terminal bulb commonly known as the cabbage. There is only one species of coconut since it is a monopypic. There are two distinct varieties of coconut, the tall and the dwarf. The varieties differ in size, shape, colour and quality of nuts, shape of crown leaves, flowers and also trunk characters. Laccadive dwarf, Andaman dwarf, Chennangi and Gangabondam in dwarf and Fiji, S.S green, San Ramon, Philippines ordinary in tall are some of the varieties found in India. The traditional west coast tall variety has a lower performance rate when compared to the exotics. Many hybrid varieties, through tall×dwarf or dwarf×tall have been crossbred in recent years with varying morphological and yield characteristics. Some combinations have shown tolerance to drought.

Being a tropical crop, coconut is cultivated normally upto an altitude of 900m in various countries. An optimum rainfall between 1300 and 23000 mm and a temperature of around 27°c (with a diurial range of 7°c) are the important climatic parameters. A warm and humid condition is preferred by the crop. Persistance of highly humid conditions may reduce transpiration thereby limiting nutrient uptake. This may favour rapid spread of diseases like budrot. Severe drought at the flowering period may kill the growing point causing the infloresence to abort. This will affect the production of nuts upto 25-30 months that follows. Therefore, rainfall during first three months of nut development becomes very essential

Sandy and clayey soil with no underlaying hardrock and proper supply of moisture and sufficient drainage are essential for coconut. The depth of the pits will depend upon the type of soil; in laterite soil with rocky substratum (as in Dakshina Kannada) deeper and wider pits have to be dug. When the water table is high, planting at the surface or even on moulds may be necessary, in which case also digging in necessary. In planting the seedlings, square to triangular systems are usually adopted. Normally 100-250 seedlings may be planted in 1 ha of land depending upon the planting system, nature of soil, rainfall and other factors. Regular manuring from the first year of planting is essential to ensure good vegetative growth, early flowering and bearing of high yields. From the fourth year onwards fertilisers have to be applied at the rates recommended for adult palms (500 g N, 320 g $P_2 0_5$ and 1200 g K_2O per palm). In addition to fertilisers, dolomite or lime and magnesium sulphate have to be applied in acidic soils. Application of green leaves and compost has to be practised for organic humus. Irrigation is practised in the semi-arid regions and also in the coasts during the off-season. Cover cropping prevents soil erosion and intercropping is also commonly practised for the various economic and ecological reasons. Different foods and other crops may be inter-cultivated in varying degrees depending upon the stage of coconut development.

The major pests of coconut are rhinoceros beetle, leaf eating caterpillar, red palm weevil, root-eating cockchafer (white grabs), slug caterpillars coried bug, etc. Mammalian pests like rats, monkeys and birds also cause damage to the nuts. The important diseases are bud rot, leaf rot, leaf blight, fruit rot, stem bleeding, anabe roga (fungus) and root (wilt) disease. Recently, Thanjavur wilt disease has become a major disease causing crop loss in many places. In a survey conducted in 1965-66, the incidence of Thanjavur wilt (garvoderma) disease was found to be 10.8 per cent in Thanjavur and between 4-6 per cent in some other districts of Tamil Nadu. During 1980-81, the incidence was between 2.6 and 13.5 per cent in eight district of that state. The wilt disease has been responsible for decay of the trees and poor yield in many of the Kerala districts. Proper plant protection measures both for prevention and cure from disease and pests are necessary. The commonly used chemicals are BHC, copper sulphate, chlordole for pests and bordeaux mixture (copper sulphate) disthane, etc. for diseases Adequate manuring (farmyard manure, green leaf, tanksilt and synthetics and application of neem oil cake control or subside the includence of pests and diseases.

Area and Production Estimates

Coconut production accounts for about 1.5 per cent of major food crop production in the world. Its share in food production in the developing countries is about 3.5 per cent. Philippines accounts for the largest area and production of coconut in the world, followed by Indonesia, India, Srilanka, Thailand and Malaysia. These countries accounted for about 86.0 per cent of the world area in 1971 and still higher than that in 1983. During 1971 the share of Philippines in world coconut area was 30.2 which increased to 32.7 per cent in 1983. In the case of Indonesia, there has been a substantial increase from 20.1 to 30.6 per cent in the period. While in India the area share actually declined from 16.4 to 12.1 per cent. (Table 3.1). In other countries also, decline in area share was noticed. Production share, however has increased in countries like India owing to increase in yield rates. During 1971 the yield of coconut was 4827 nuts per ha for all the countries combined but the level declined to 3890 nuts in 1983, showing an average annual decline of about 1.6 per cent. The yield decline was very marked in Indonesia (from 6110 to 2874 nuts), Philippines (from 4222 to 4357 nuts) and Malaysia (from 4694 to 3972 nuts), whereas Srilanka (from 4212 to 4735 nuts), Thailand (from 1375 to 2258 nuts) and India (from 5572 to 5772 nuts) registered increase over the base year yield levels. However, the yield level in India as would be shown latter has shown some decline in various states.

According to Indian statistics, coconut area increased from 0.99 mil. ha 1965-66 to 1.16 mil ha during 1983-84 while the yield rate declined from 5700 to 5100 nuts per ha during this period. Upto 1970-71 the yield level has been increasing and since then showing a decline. Among the different states, Kerala is the major coconut producer. The yield decline has been very substantial in this state declining from 5500 nuts/ha, in 1970-71 to just 4000 nuts/ha in 1983-84. In Andhra Pradesh, similar decline in yield rate is noticed. In terms of area share, Kerala accounted for 66.3 per cent in 1965-66, 68.8 per cent in 1970-71 but the share declined to 66.7 per cent in 1975-76 and further down to 58.6 per cent in 1983-84. Yield decline in states like Kerala was responsible for the national average to decline. Among the coconut producing states, Tamil Nadu maintains a stable yield level (between 9300 to 10200 nuts/ha though in West Bengal, Pondichery and Lakshadweep, the current yield level is above the national average. In case of Karnataka, the area share slightly declined from 13.1 per cent 1965-66 to 12.5 per cent in 1970-71 and from thereon increased to 13.2 per cent 1975-76 and further upto 16.1 per cent in 1983-84. All the states except Kerala in area and Kerala and Andhra Pradesh in yield have shown increased share. Table 3.3, shows the area share by different states in different years, showing improved status for Karnataka and Tamil Nadu and a declined one for Kerala.

Table 3.4 presents the compound growth rates of area, production and yield of coconut durıng 1951-52 to 1980-81 and 1966-67 to 1980-81 in respect of the major states. In area, the growth rate has been remarkable in Karnataka, Tamil Nadu and Orissa. This is the case with production also, Maharashtra included. The yield growth rate has been negative during 1951-52 to 1980-81 in India declining at the rate of 0.82 per cent per annum. The rate of decline was slightly lower during 1966-67 to 1980-81 indicating a larger decline in the pre-1966-67 era. In Kerala and Andhra Pradesh negative growth rate has been very high (1.34 and 3.38 per cent respectively for the entire period).

According to Table 3.5, in Kerala coconut accounted for 31.02 per cent of net sown area in 1983-84, increasing from 23.69 per cent in 1961-62. The peak was during 1971-72 period when the proportion stood at 33.39 per cent. In Karnataka coconut share increased from 0.97 to 1.75 per cent in the course of 32 years. In Tamil Nadu also a similar marked increase in the area coverage (from 1.27 to 2.4 per cent in the period) is seen. In the country, the area share of coconut is very marginal (increasing from 0.53 to 0.81) but the rate of increase has been quite substantial.

Coconut Cultivation in Karnataka

In Karnataka coconut cultivation is confined to the coastal districts to Uttar and Dakshina Kannadas, the malnad region of Chickmagalur, Hassan, Kodagu and Shimoga and southern plains of Bangalore, Chitradurga, Kolar, Mandya, Mysore and Tumkur and the northern plains of Bellary, Dharwad and Raichur. In other districts also the crop is cultivated to a small extent. Tumkur district leads in area and production followed by Hassan, Chickmagalur, Chitradurga and others. The above five districts accounts for over three fourths of coconut area in the state.

Table 3.6 presents the share of coconut by different districts in Karnataka; coconut in the state increased from 0.89 lakh ha in 1955-56 to 1.86 lakh ha in 1983-84. Coconut is one of the few crops where a substantial increasing area has been noticed. The crop is grown in most of the district over the thirty years some changes in area share by the districts have taken place. During 1955-56, Tumkur, Hassan and Dakshina Kannada shared about 63.3 per cent of total coconut area in the state during 1983-84. The proportion came down to 53.3 per cent in 1983-84. Tumkur has slided in its share (from 29.7 to 24.2 per cent) though it still leads over others. There has been a decline in a case of Dakshina Kannada from 13.3 to 9.0 per cent. Hassan, Bangalore and Mysore maintained their respective shares. The coastal districts have been the major losers. Chickmagalur, Chitradurga, Mandya and other districts registered in increase in area share.

Table 3.7 describes the percentage of area under coconut to gross cropped area in the different districts. It could be seen that in all the districts there has been an increased share. For the state it increased has been from 0.85 per cent in 1955-56 to 1.6 per cent in 1980-81. In Mandya, Chitraduga, Shimoga and in the non-traditional areas, the increase in coconut area has been very substantial. In the coastal districts, Hassan and

Tumkur the increase has been quite marked. In Tumkur, the proportionate increase was from 5.91 to 8.23 per cent, in Hassan from 6.66 to 8.66 per cent, in Chickmagalur from 4.34 to 7.48 per cent, in Dakshina Kannada from 4.55 to 5 8 per cent and in Uttara Kannada from 3.95 to 4.18 per cent. The other districts where coconut share is marked are Chitradurga, Mandya, Bangalore and Mysore.

Table 3.8 presents yield level of coconut in the districts. The yield increased from 4011 nuts/ha in 1955-56 to 5190 nuts/ha in 1983-84 for the state as a whole. This amounts to about 1.0 per cent growth in the yield level per annum. In 1955-56, the yield level was very high (over 9000 nuts/ha) in Mandya, Kodagu and Uttara Kannada; in 1983-84, only Tumkur has the highest yield rate at 6871 nuts/ha, Hassan, Dharwad, Raichur, Mandya, Bangalore, Chitradurga, Kolar and Kodagu report over 5000 nuts/ha. Over the years, the yield level has declined substantially in Uttara Kannada, Chickmagalur, Kodagu and Mandya. Hassan, Shimoga, Chitradurga, Tumkur, Bellary, Dharwad and Raichur have registered a substantial increase in the yield level. Yield level increase was marginal in Dakshina Kannada, Kolar and Mysore. Only in Chickmagalur that the yield level has shown a decline. On the whole the major coconut producing districts have registered a positive growth in area, production and yield.

Table 3.9 describes the compound growth rates of area production and yield of coconut during 1955-76 in Karnataka. It is seen that the area has increased at 2.6 per cent per annum while production registered a growth of 3.6 per cent, indicating growth rate of 0 9 per cent in yield level. During this period, the gross cropped area in the state has increased at the rate of 0.2 per cent per annum. But in production and yield rate, coconut has registered only a moderate increase when compared to maize, wheat, sugarcane and arecanut where the annual increase has been from 6.0 to 26.0 per cent in production and 2.6 to 7.2 per cent in yield rate. Estimation of yield in coconut is subject to many approximations and manipulations. Hence the yield level has to be considered with due limitations. The Food and Agriculture Organisation Yearbook presents the yield level in metric tonnes approximating 1 kg for one coconut.

The districts where the area growth has been very substantial are all except the coastal districts; in production also, the coastal districts have performed negatively. The negative yield rate is encountered in coastal districts, Chickmagalur, Kodagu, Mangalore, and Kolar increased in the yield rate has been marked in the northern plains and in Mandya, Chitradugra, Tumkur, Hassan and Shimoga.

The importance of coconut in state economy could be highlighted taking into consideration its share in agricultural value in different years. Whereas the share in area was only 0.85 per cent 1955-56, coconut had contributed about 3.41 per cent of gross agricultural value in the state. But in the subsequent period, the value share did not increase correspondingly to increasing area share. During 1979-80 the contribution of coconut to gross agricultural nature stood at 4.1 per cent for an area share of about 1.6 (Table 3.10).

Districtwise, the share has remained more or less constant in Dakshina Kannada, Kolar, Mysore and Kodagu. Decline in contribution is noticed in Uttara Kannada (from 14.18 to 5.64 per cent), Chickmagalur (6.71 to 4.26 per cent), Bangalore (4.21 to 2.95 per cent) and Mandya (8.45 to 3.08 per cent). The districts where importance of coconut has gone up are Tumkur (17.2 to 22.09 per cent), Hassan (12.33 to 17.02 per cent), Chitradurga (from 4.7 to 7.58 per cent), Shimoga (0.16 to 0.43 per cent and the Northern plains. It is seen that Tumkur, Hassan and Chitradurga have been the major districts were coconut performance has been very much encouraging if the combination of area, production, yield and gross value contribution is considered.

TABLE 3.1: **Estimated world area and production of coconut**

	Area[1] ('000/ha)	1971 Production ('000/MT)	Yield mt/ha	1981 Production[2] ('000/MT)	1981 Copra-production ('000/MT)
W orld	5949	29355	4.93	36665	5054
Philippines	1800	7601	4.22	11050	2275
Indonesia	1200	7333	6.11	10800	1254
India	978	4472	4.57	4500	370
Sri Lanka	466	1963	4.21	1716	123
Malaysia	272	1277	4.69	1207	208
Oceania	316	2187	6 92	2274	105
Others	917	4522	4.93	5118	719

Sources : 1. R. Child, *Coconuts*, Longman, London, 1964. p. 30.
2. *FAO Production Year Book*, 1981 (1969-71 average for 1971)

TABLE 3.2: **Area and Yield of Coconut in Indian States**

	Area ('000/ha)				*Yield ('000 nuts/ha)*			
	1965-66	*1970-71*	*1975-76*	*1983-84*	*1965-66*	*1970-71*	*1975-76*	*1983-84*
India	983.7	1045.7	1125.4	1154.8	5.7	5.8	5.4	5.1
Kerala	586.3	719.1	751.1	676.4	5.4	5.5	5.0	4.0
Karnataka	115.4	130.3	148.3	186.2	3.1	5.6	6.0	5.2
Tamil Nadu	90.2	95.1	110.3	140.6	9.3	9.9	10.2	9.8
Andhra Pradesh	33.3	37.3	40.6	46.5	8.6	4.3	4.1	4.2
Goa+	19.7	19.7	18.7	21.6	3.6	3.6	6.5	4.9
Andamans	8.6	8.7	19.7	20.8	4.4	4.3	3.3	4.6
Maharashtra	8.2	9.6	9.1	11.2	3.9	4.9	6.0	5.5
Orissa	7.5	10.6	11.6	23.3	5.8	3.7	3.8	4 2
West Bengal	6.7	6.7	6.7	14.9	3.3	3.3	3.3	14.2
Assam	3.6	3.8	4.2	7.5	3.1	3.0	3.5	6.4
Lakshadweep	2.7	2.8	2.8	2.8	7.1	7.0	7.6	7.9
Pondicherry	1.4	1.6	1.6	1.6	9.9	10.0	10.2	8.8
Tripura	0.2	0.4	1.0	1.4	1.1	1.4	1.3	1.2

Source : Agricultural situation in India, Various issues.

TABLE 3.3: **Area share in Coconut by different states**

State	*1965-66*	*1970-71*	*1975-76*	*1983-84*
Kerala	66.3	68.8	66.7	58.6
Karnataka	13.1	12.5	13.2	16.1
Tamil Nadu	10.2	9.1	9.8	12.2
Andhra Pradesh	3.7	3.6	3.6	4.0
Goa	2.2	1.9	1.7	1.9
Andamans	1.0	0.8	1.8	1.8
Maharashtra	1.0	0.9	0.8	1.0
Orissa	1.0	1.0	1.0	2.0
Others	1.5	1.4	1.4	2.4

TABLE 3.4: **Compound growth rates of Area, Production and Yield of coconut**

(Percentage p.a.)

State		*1951-52 to 1980-81*	*1966-67 to 1980-81*
Kerala	A	2.10	0.10
	P	0.73	—1.65
	Y	—1.34	—1.74
Karnataka	A	2.56	3.36
	P	3.42	4.93
	Y	0.84	1.53
Tamil Nadu	A	3.41	2.25
	P	4.33	2.62
	Y	0.89	0.36
Andhra Pradesh	A	0.73	1.41
	P	—2.67	—1.17
	Y	—3.38	—2.56
Maharashtra	A	0.95	0.43
	P	2.31	3.80
	Y	1.34	3.36
Orissa	A	5.54	6.82
	P	2.19	8.65
	Y	—3.17	1.72
India	A	2.40	0.98
	P	1.56	0.20
	Y	-0.82	—0.77

Source: P.K. Das, 'Coconut Situation in India', *Agricultural Situation in India*, August 1983, p. 280.

TABLE 3.5: **Percentage of area under coconut to net sown area**

State	*1951-52*	*1971-72*	*1976-77*	*1983-84*
Kerala	23.69	33.39	31.58	31.02
Karnataka	0.97	1.28	1.67	1.75
Tamil Nadu	1.27	1.80	1.81	2.40
Andhra Pradesh	0.33	0.34	0.37	0.41
Maharashtra	NA	0.06	0.05	0.06
Orissa	0.08	0.17	0.22	0.39
India	0.53	0.78	0.77	0.81

TABLE 3.6 : Share of coconut area by different Karnataka Districts

(Per cent)

District	*1955-56*	*1960-61*	*1965-66*	*1970-71*	*1975-76*	*1983-84*
Dakshina Kannada	13.3	11.3	10.6	9.1	8.9	9.0
Uttara Kannada	5.4	4.6	4.2	3.6	3.3	2.7
Chickmagalur	9.4	9.6	8.9	12.8	12.3	11.0
Hassan	20.3	22.0	22.0	20.8	19.3	20.1
Kodagu	0.1	0.1	0.1	0.4	0.4	0.3
Shimoga	0.3	0.3	0.3	0.5	1.0	1.8
Bangalore	4.0	3.0	2.8	3.4	3.7	4.7
Chitradurga	7.0	9.7	12.3	10.9	10.6	12.6
Kolar	0.5	0.4	0.4	0.4	0.5	0.5
Mandya	3.9	4.6	4.2	5.5	5.5	5.8
Mysore	5.2	5.1	5.2	5.0	5.9	24.2
Bellary	0.6	0.5	0.4	0.8	0.3	0.3
Dharwad	0.2	0.2	0.2	0.2	0.2	0.2
Raichur	—	—	—	—	—	0.1
State (000 ha)	88.61	104.49	114.33	133.16	151.52	186.22

TABLE 3.7 : **Percentage of area under coconut to total cropped area in Karnataka**

District	*1955-56*	*1960-61*	*1965-66*	*1970-71*	*1975-76*	*1980-81*
Dakshina Kannada	4.55	4.26	4.21	4.47	5.08	5.80
Uttara Kannada	3.95	3.64	3.55	4.20	4.15	4.18
Chickmagalur	4.34	4.83	4.60	7.04	7.20	7.48
Hassan	6.66	8.01	5.72	8.20	7.75	8.66
Kodagu	0.10	0.03	0.11	0.41	0.37	0.47
Shimoga	0.14	0.10	0.12	0.18	0.46	0.73
Bangalore	1.01	0.87	0.72	1.13	1.37	1.75
Chitradurga	1.14	1.88	3.20	2.63	2.78	3.50
Kolar	0.14	0.14	0.11	0.14	0.21	0.17
Mandya	1.55	1,88	1.85	2.50	2.50	3.39
Mysore	1.04	1.12	1.81	1.23	1.48	1.73
Tumkur	5.91	5.82	5.28	7.16	7.38	8.23
Bellary	0.08	0.07	0 07	0.07	0.07	0.10
Dharwad	0.01	0.01	0.02	0.02	0.03	0.04
Raichur	—	—	—	—	—	0.02
State	0.85	0.99	1.10	1.22	1.36	1.60

TABLE 3.8 : **Yield of Coconut (nuts/ha) in Karnataka districts**

District	*1955-56*	*1960-61*	*1965-66*	*1970-71*	*1975-76*	*1983-84*
Dakshina Kannada	3689	4623	4508	3883	4328	4330
Uttara Kannada	9439	8016	3826	4129	4865	4855
Chikmagalur	3282	6813	1531	4871	2397	2397
Hassan	2604	4344	3355	6021	5506	5506
Kodagu	11393	9594	5273	5418	3752	5034
Shimoga	1971	1789	5333	1500	3291	3289
Bangalore	5436	3590	4828	3889	3752	5976
Chitradurga	2960	3435	2368	6951	5791	5827
Kolar	4362	4198	5196	3893	3752	5056
Mandya	11434	4554	5177	5417	3752	5074
Mysore	3027	4179	1566	6895	3357	3346
Tumkur	3707	4668	4045	3232	6871	6871
Bellary	751	7742	5165	4847	3758	5074
Dharwad	874	4877	5213	4136	3754	5501
Raichur	1000	—	—	5143	3750	5216
State	4011	4774	3434	4852	5075	5190

Source : Directorate of Economics and Statistics, Bangalore.

TABLE 3.9 : **Compound growth rate of area, production and yield of Coconut in Karnataka, 1955-76.**

District	*Area*	*Production*	*Yield*
Dakshina Kannada	0.40	—0.96	—1.37
Uttara Kannada	0.10	—1.51	—1.36
Chikmagalur	4.30	1.80	—2.39
Hassan	2.30	1.80	—2.90
Kodagu	14.10	13.30	—0.58
Shimoga	7.90	11.50	3.30
Bangalore	3.30	2.80	—0.44
Chitradurga	4.00	5.80	1.20
Kolar	4.20	4.00	—0.26
Mandya	4.60	6.40	1.70
Mysore	3.40	4.30	0.90
Tumkur	2.20	4.80	2.60
Bellary	0.60	2.20	1.60
Dharwad	3.80	7.40	3.50
Raichur	8.80	12.90	4.50
State	2.60	3.60	0.90

TABLE 3.10 : **Gross value share of Coconut in Karnataka agriculture.**

(Per cent) (at 1970-73 average State farm harvest prices)

District	*1955-56*	*1960-61*	*1965-66*	*1970-71*	*1975-76*
Dakshina Kannada	7.44	9.19	9.55	6.87	7.12
Uttara Kannada	14.18	11.71	5.52	5.40	5.64
Chikmagalur	6.71	9.10	3.31	9.23	4.26
Hassan	12.33	18.76	16.13	21.60	17.02
Kodagu	0.35	0.05	0.14	0.37	0.28
Shimoga	0.16	0.09	0.33	0.09	0.43
Bangalore	4.25	2.80	5.87	3.11	2.95
Chitradurga	4.70	7.21	8.82	10,28	7.58
Kolar	0.47	0.49	1.13	0.33	0.43
Mandya	8.45	4.27	4.06	5.19	3.08
Mysore	2.80	3.85	2.19	4.89	2.60
Tumkur	17.20	19.37	30.44	14.59	22.09
Bellary	0.05	0.61	0.41	0.20	0.18
Dharwad	0.01	0.07	0.15	0.07	0.09
State	3.41	4.36	3.69	3.79	3.92

Source : M. Prahladachar and S.M. Sundararaju, *op. cit.*

4

Coconut Cultivation in Dakshina Kannada

Introduction

Coconut is essentially a small-holder-crop unlike other commercial crops. It is grown mostly in homestead gardens and small holdings. There are about 5 mil coconut holdings in the country. Holdings of less than 2 ha account for 98.0 per cent of total holdings. Kerala alone accounts for 50.0 per cent of the total cocount holdings in the country. It is estimated that about 10 million people depend directly or indirectly on coconut cultivation and industry for their livelihood. In Kerala about 15.0 per cent of the state income and 30.0 per cent of agriculture income are countributed by coconut. At the national level the average value of production of coconut was estimated to be around Rs. 900 cr with an export earning of Rs. 26 cr normally through coir and coir products.

In Karnataka, coconut holdings of less than 0.2 ha account for about 52.5 per cent of total holdings, while the proportion of 0.2-1 0 ha holdings is 42.0 per cent. Holdings of 1 ha account for about 4.6 per cent according to a survey conducted by then State Bureau of Economics and Statistics. During 1972-73, pure type of coconut gardens accounted for 61.6 per cent, the rest being accounted for by the mixed types. In Dakshina Kannada, the pure types shared 76.6 per cent of gardens, while the proportion was 85.9 per cent in Hassan, 71.2 per cent in Tumkur, 58.1 per cent in Chitradurga and a mere 9.8 per cent in Uttara Kannada. In the mixed type gardens, coconut was mixed with arecanut (22.4 per cent), mango (1.0 per cent), banana (6.4 per cent), areca+banana+mango (58.6 per cent), betalvines (0.2 per cent) and others including cereals, pulses, tamarind and jack. In Dakshina Kannada, coconut is mixed mainly with arecanut (71.9 per cent), paddy and spices (15.8 per cent), banana (11.3 per cent) and mango (2.3 per cent). The districts where coconut is mixed mainly with arecanut are Hassan and Tumkur, those with arecanut+banana+mango being Uttara Kannada, Chikmagalur, Mysore and Tumkur. It is mixed with cereals in Chitradurga and to a certain extent in Mysore.

Coconuts are plucked mainly in the months of January to June (about two third of harvest will be completed in this period) and the keeping quality of copra will be good when nuts are plucked during March to June. Coconut usually ripens in about 12 to 13 months after the opening of the inflorescence. In order to get the maximum yield of copra and oil only fully matured nuts should be harvested. The loss of copra is 6.0 per cent in 11 months old nut, and 16.0 per cent in 9 months old nuts. The reduction in the percentage of oil in the 11, 10 and 9 months old nuts is 5, 15 and 33 per cent respectively. In the west coast, 6 to 12 harvests a year are usual practices. In many areas of Kerala, harvesting is done at an interval of 45 days during summer and 60 days during the rainy season.

The yield of coconut plam is influenced by agro-climate, as well as cultural practices. The tall varieties produce on an average 60-80 nuts per palm per year. There are cases of trees yielding more than 100 nuts in the case of some Laccadive varieties. Even among the dwarf varieties, a mean yield of 70-120 nuts is a common experience. Depending upon the various yield capacity, the trees have been grouped into heavy yielders (more than 80 nuts per year), medium yielders (41-80 nuts per year) and poor yielders (40 nuts and below per year). Usually the proportion of heavy yielders will be small in any coconut garden. In countries like India the hot summer season accounts for over a third of the annual crop. Though the per ha yield of coconut is comparatively high in India, the per palm yield of nuts and the unit output of copra are low. While the average per palm productivity is 35 nuts in India, it is 62 in Western Samoa, 45 nuts in Sri Lanka and Malaysia and Philippines average the Indian productivity at 35 to 40 nuts. In other countries 4500-5000 nuts and 20000 or even less whole shells are required to produce 1 tonne of copra and charcoal respectively, but the corresponding figures for India are 6800 nuts and 30000 or more whole shells. Similarly, Sri Lanka and Philippines produce nearly 25.0 per cent more fibre from a unit quantity of husk compared to India. The major factor responsible for these is the characteristics of the cultiver west coast tall grown in India producing comparatively small size nuts with low copra output. Coconut can be an irrigated crop also and in Karnataka about 30.0 per cent of coconut area is irrigated. The average for the state is 30.0 per cent, the proportion being very high for D.K, U.K, Chikmagalur and Chitradurga and very low in Tumkur and Hassan. The major source of irrigation is by wells (73.3 per cent). The distribution of coconut garden by number of palms shows that about 44.6 per cent of the gardens have less than 25 palms per garden, palms 28.4 per cent between 25-49 palms, 13.5 per cent between 50-74 palms, 6.0 per cent between 75-99 palms and 8.5 per cent over 100 palms. At the extreme 1.4 per cent of the gardens have more than 300 palms per garden.

According to 1983-84 estimates the per palm productivity in Kerala is only 25 nuts as against 44 nuts in Tamil Nadu and Lakshadweep and 54 nuts in Karnataka. Irrigation is not a common practice in Kerala and where irrigation is practised, productivity is very high.

Coconut Cultivation in D.K.

The area under coconut in the district increased from 11,778 ha in 1955-56 to 12,157 ha in 1965-66, 13,466 ha in 1975-76 and 18,014 ha in 1985-86. On an average, the area has been increasing at an annual rate of 1.7 per cent. The yield per ha has increased from 3.689 nuts in 1955-56 to 4,623 nuts in 1960-61 after which there was a slight decline. During 1970-71 the yield level declined to 3,883 nuts and during the laste decade it has been stabilised around 4,330 nuts. Talukwise, Udupi (4,763 ha), Kundapur (3,026 ha), Karkal (2,089 ha) and

Bantwal (2,071 ha) are the major coconut producing taluks. Belthangady (1,924 ha), Mangalore (1,776 ha) and Puttur (1,505 ha) also report substantial area under coconut. (Table 4.1). In percentage terms coconut occupies about 6.0 per cent of the gross cropped area in the district during 1985-86. The highest proportion is in Udupi (8.7 per cent) and Kundapur (6.8 per cent). In Puttur, Bantwal, Mangalore and Karkal the share is between 5.6 to 5.1 per cent. In Belthangady and Sullia the proportion is more than 4.0 per cent. Among the taluks Puttur represents the district average.

Age and Variety

Coconut cultivation in the two different categories (with or without paddy) is considered as to the age of the palms and type of variety. The local variety is called west coast tall and the hybrid variety a cross between dwarf and tall or tall and dwarf is also cultivated recently. The keeping quality of the hybrid coconut and also copra is poor when compared to that of the locals. Because of this the yield is usually disposed of as tender coconut or coconut for table purpose.

The proportion of hybrid varieties was 6.3 per cent for the entire sample In the first category the average worked out to about 7.0 per cent. The lower size groups fared slightly better in area of hybrid varieties. This was the case in the second category also. For the hybrid varieties, the proportion of pure type gardens was 35.1 per cent, the proportion being higher (46.2 per cent) in the first category. The palms were usually of less than 7 years old though in the fourth group. In the first category, about 25.0 per cent of the hybrid palms belong to the 7-15 years age group. The average yield of hybrid palms was around 34 nuts per palm. There was little difference in the total number of trees per ha (around 175). The yield of hybrid palms was about 7.0 per cent higher than that of local varieties.

All the farmers grow the local varieties and the hybrids and dwarfs occupy more or less an ornamental place. Though the yield in the initial year may be very high (over 100 per palm) in the hybrids, in latter years the yield would decline. Because of proven stability in yield and quality, the locals will remain the major varieties for long. In the sample over 93.0 per cent of the coconut area was under the local varieties. The local trees live longer than the hybrids.

The local trees have been grouped under less than 7 years, 7-15 years, 15-30 years and 30 years and above. In the sample area the proportion of pure type of garden was very long unlike the district average (especially as in Udupi, Kundapur and Karkal). Table 4.2 presents particulars about age, type of garden, etc. In the first category, the proportion of pure type of gardens was 9.7 per cent; it was 5.3 per cent for less than 7 year old trees, in that category. It was 3.9 per cent in the 7-15 years and 4.4 per cent in the 15-30 years old groups. Trees of above 30 years were in the mixed type gardens. The less than 1 ha group did not possess any pure garden. The proportion of pure garden in the second category was slightly higher at 11.8 per cent. Here also the older trees (above 30 years) and the first two size groups had not pure gardens. The proportion was about 4.0 per cent each in the other age groups. It may be seen that the proportion of pure type of garden was lower in the case of local varieties. Since the garden plots are very small whether in the homestead or farmstead, the notion of pure type was subjected to local manipulations like area, lacality, crops to be mixed and number of coconut trees.

The trees of less than 15 years accounted for 49.8 per cent in the first category and 55.9 per cent in the second. In the first category trees between 15-30 years shared 33.4 per cent of total population. The proportion was quite substantial in the first and fourth groups. In the second category also they dominated though trees between 7-15 years shared a substantial proportion (34.8 per cent as against 35.9 per cent in the succeeding age group). The number of old trees were very high in the second category.

Cost of Cultivation

An attempt has been made to compute the cost of cultivation per ha coconut garden. Coconut is a small holder crop.

During 1972-73, coconut gardens of less than 25 palms per garden accounted for about 45.0 per cent of total number of gardens. Gardens between 25 to 49 palms accounted for over 28.0 per cent and between 50 to 74 palms 14.0 per cent. Hence computation of cost per ha may not be representative of a typical type of coconut garden. The cultural practices followed in a garden of 5 to 10 palms would be entirely different from that having more than 50 palms. The benefit cost ratio would be different for different years, varieties and method of irrigation. Though material cost and price of coconut may not vary between different regions or farmers, labour wages may show varying levels in different situations. The initial investment in land development, seedlings and planting will be very high. Depending upon the variety the initial investment and the recurring expenditure, benefit will be realised from the 8th or 12th year of plant. Various cost and return figures are available. According to the coconut pamplet published by the Central Plantation Crops Research Institute. The westcoast tall variety can give a net profit of at least Rs. 14,000 from the 13th year and the D×T hybrid at lesast Rs. 25,000 from 10th year taking into consideration yield and price fluctuations. The net benefit cost ratio would work out to 0.04 for WCT and 1.47 for D×T varieties in the 7th year. The ratio would be 1.23 for WCT and 2.58 for the D×T hybrid varieties in the 9th year and 2.33 and 4.1 respectively in the 15th year.

According to another study by the same Institute the net benefit cost ratio would be 0.12 in the first bearing year ; the ratio would increase to 1.23 in the third bearing year, 2.32 in the fourth bearing year and 2.37 from the 11th year onwards. It is evident that from coconut cultivation the initial investment and the annual expenditure can be realised even in the earlier stages of bearing and a very favourable benefit cost ratio realisable after the fifteenth year.

The major items of cultivation expenditure include land preparation, manure and fertiliser application, planting of seedlings, basin formation, watering and interculture in the first year. In the subsequent years, gap filling, basin formation, watering, manure and fertiliser application, interculture and plant protection would be the different operations to be undertaken. Land preparation and planting consume the bulk of expenditure. Plant protection (including watch and ward) is a major cost item in the initial year. In the sample the cost of cultivation particulars referred to a point of time irrespective of different age groups of palms. Though the farmers had reported different age-groups over 70.0 per cent fall in the 7-30 age-group) the cost particulars referred to the average. The different size groups and the coconut size groups therein showed positive relationship between the two. As the general size group increased, coconut area also increased. The cost and returns particulars in respect of different size groups averaged out the performance between differrnt coconut size groups and age-groups.

The cost particulars included labour cost (both hired and imputed family labour) expenses on manure and fertilisers, power cost and miscellaneous. Labour cost would include expenditure on land preparation, interculture, watering, plant protection, harvesting, etc. Power cost would include expenses on electricity or diesel to run the irrigation pumpsets, tractors, etc The miscellaneous cost would include other expenses, imputed interest, land revenue, etc., (about 20.0 per cent total cost). However, land value appreciation has not been taken into cost calculation.

In the first category, labour requirement was higher in the case of hired labour per ha. The requirement of hired labour was 38.1 mandays while that of family labour the quantity was 14.9 mandays. The average wages per manday has been Rs. 15 per diem. This included kind wages in particular season. In the second category, the proportion of family labour was equal to that of hired labour unlike in the first category. Labour requirement was 17.0 per cent higher here. Because paddy was cultivated in the first category, bulk of family labour was diverted to its cultivation. However, in the second category owing to absence of paddy cultivation a higher proportion of family labour was allocated to cocount cultivation. On average total labour requirement per ha of coconut cultivation (for both categories) was 56.3 mandays of which about 63.0 per cent was accounted for by hired labour.

Animal ploughing was not resorted to in coconut cultivation. Land development, basin formation and interculture practices were done by manual labour both in the pure or mixed type gardens. Organic manure in the form of cattle dung or compost was been applied to the palms. Besides, green manure during the rainy season and ash after the monsoon were also applied. Chemical fertilisers (usually NPK complex) were applied but about 58.0 per cent of farmers did not apply any fertilisers. The proportion was higher in the first category (64.0 per cent) owing to the fact that in the mixed type of gardens, fertilisers applied to paddy or arecanut might be utilised by coconut also. Chemical fertilising was still partial ; about 45.3 kgs of chemical fertilisers have been applied to an ha of coconut plantation. If fertiliser application was considered for the applied area only the per ha consumption would be 107 kgs. In manure consumption not much difference was noticed between the two categories. Though the first size group in second category reported a higher dose of consumption, the average fertiliser consumption was relatively higher in the first category (especially in II and IV groups). In the second category, the first size group did not consume any chemical fertilisers. Even in the first category the first group consumed only 15 kgs per ha. Other material inputs like electricity, diesel were included in the expenditure to the extent applicable to the particular crop.

Table 4.3 presents total cost of cultivation which included labour cost, (wages for family labour imputed at the market rate) expenditure on organic manure and chemical fertilisers, electricity and diesel and other expenses. Since the cost did not take into consideration the different age-groups or varieties it was meant to be an average only. The cost did not represent the spread-over average for the first 7 years or 15 years and the like. The column under the cost includes rental value, interest spread over initial cost, plant protection, expenses and miscellaneous items of expenditure for the particular farm concerned endowed with different varieties and age-groups. Taking all this into due consideration net benefit cost ratio was 2.71 for the entire sample, lower at 2.34 in the first category and 3.37 in the second. For calculating the income, harvest prices have been taken into consideration around Rs. 2 per nut) and income from by products whether sold or consumed domestically was not included. Along with coconut some minor crops were raised and the average income per household amounted to Rs. 280, in the first category Rs. 257 and in the second Rs. 320. In the first category, only the second group had subsidiary income by coconut. In the second category except the first group all others had subsidiary income ranging between Rs. 200-500.

The first category had a lower BCR and it was lower than the average in the fourth group. In second category, the third group had a lower ratio, while the first reported a higher ratio. The point to be made here mainly concerns with the positive and favourable cultivation, scenario of byproduct cultivation and the components.

Of the lotal coconut production in the sample about 6.0 per cent in the first category and 3.2 per cent in the second was disposed of in the form of tender coconuts. The proportion was higher in the first groups of the first category. About 26.2 and 26.5 per cent is home consumed. The proportion was substantial in the first two group of the first category and first group of second category. Bulk of coconut produce was sold out in the form raw coconut and copra both together sharing 67.8 per cent in the first category and 80.3 per cent in the second category. The proportion of coconut sale and copra was comparatively higher in the latter category. The bigger size groups had a larger sales proportion.

Coconut has many byproducts, the average dry biomass of coconut is estimated to be over 10 t/ha. All the parts of the plant are usuable. Coconut fronds, leaves, tetioles, hunch, stalks, spathes, spipules, etc, are used for thatching, fencing basket-making, broom-stick making and as fuel. Coconut shells (tender corp) is used as fuel, hookah, shells, curios and activated carbon. Coconut husk is used as fuel and in manufacture of coir and coir products. The pith is used as manure or as a basestock to be mixed with earth or lime.

In the sample area coconut byproducts were used for domestic fuel consumption and for manure purposes. Sometimes coconut fronds were sold for thatching purpose. Some of the tile factories and ironsmiths bought coconut shells for fuel purposes. Breaking and making the shells into small chips and selling attracts demand from town people for water heating and other purposes.

Coconut is used in the dessicated form for culinary and confectionary purposes. Karnataka is favoured for production of dessicated coconut. Increased production cf this will reduce domestic demand for copra such that copra for oil extraction will increase reducing import of oil.

Coconut oil shares the plight of oilseed economy in the country. The poor performance of the oilseeds sector has necessitated import of oil and fat in large quantity. Indigeneous coconut oil production has shown only a marginal increase. The production increased from 1.72 lakh t in 1975-76 to 2.18 lakh t in 1983-84. The major reason for stagnation in coconut oil production has been the poor performance of Kerala where 95.0 per cent of the copra is milled. Starting from the 70s the high incidence of root wilt disease to the extent of 40.0 per cent of total coconut area in Kerala reduced coconut production by over 26.0 per cent. Besides this disease, lack of replanting due to palm senility and underplanting in addition to poor rainfall in recent years have been responsible for declining coconut growth.

Coconut oil is used for edible and inedible uses. The edible portion amounts to 38.0 per cent and the rest 62.0 per cent is used for industrial uses (soap making, paints, pharmaceuticals, etc). The per capita consumption of oils and fat in India is about 6.7 kgs as against 8 kgs in the developing countries and over 13 kgs for the world. Owing to oilseeds stagnation, import of oil (coconut oil from Philippines and palm oil) becomes essential to maintain a stable oil consumption. As in the case of pulses where per capita consumption has actually declined, a similar tendency confronts the oilseeds industry also.

Control of pest and diseases to protect the plant becomes a crucial factor in increasing yield in the incidence-prone states like Kerala. In the sample area no major diseases or pests were encountered by the farmers in the recent past. The incidence of the wilt disease is very negligible. Spraying with pesticides or fungicides is a rare phenomenon in the sample area. Whenever minor pests or diseases are encountered they are treated as natural or some biological control effected manually. Chemical control was practised in extreme cases.

Coconut is spreading in the non-traditional areas at both macro and micro levels. About 0.5 mil ha is planned to be brought under coconut cultivation in the non-traditional states like Bihar, Madhya Pradesh, Orissa, West Bengal, etc. At the micro level coconut is being introduced in mulberry plots, paddy fields and even in the slopy regions. Its importance has a border crop has enabled escalation of its area in many places.

TABLE 4.1 : **Coconut area in different Taluks of D.K (in ha)**

Taluk	*1983-84*	*1985-86*	*Proportion of GCA under coconut (%)*
Bantwal	1810	2071	5.5
Belthangady	1800	1924	4.9
Karkal	2021	2089	5.1
Kundapur	2694	3026	6.8
Mangalore	1720	1776	5.4
Puttur	1453	1505	5.6
Sullia	886	860	4.2
Udupi	4737	4763	8.7
Total	17121	18014	6.0

TABLE 4.2 : **Age and Variety in Coconut (per cent)**

Size Group	*Less than 7 years*				*7-15 years*			*15-30 years*		*30+years*	*Total No. of trees*			
	Local		*Hybrid*		*Local*		*Hybrid*				*Local*		*Hybrid*	
	Pure	*Mixed*	*Pure*	*Mixed*	*Pure*	*Mixed*	*Pure*	*Pure*	*Mixed*	*Mixed*	*Pure*	*Mixed*	*Pure*	*Mixed*
Category I														
1	—	12.5	—	100.0	—	25.0	—	—	50.0	12.5	—	40.0	—	5.0
2	3.1	15.3	88.2	11.8	3.1	47.8	—	9.2	18.4	3.1	6.25	34.4	3.75	0.5
3	9.9	9.9	18.5	81.5	8.7	32.9	—	6.2	29.2	3.2	33.3	100.9	1.7	7.5
4	4.5	10.6	75.0	—	—	27.3	25.0	—	45.5	12.1	7.5	157.5	10.0	—
Average	5.2	11.9	46.0	48.1	3.8	34.1	5.8	4.5	33.7	6.8	14.7	86.3	3.8	3.6
Category II														
1	—	16.6	—	—	—	50.0	—	—	33.4	—	—	30.0	—	—
2	—	12.3	—	100.0	—	41.1	—	—	41.1	5.5	—	91.25	—	6.25
3	0.6	15.2	41.7	58.3	5.5	29.9	—	13.4	32.9	2.5	40.0	165.0	6.25	8.75
4	13.1	9.6	12.5	87.5	7.2	34.7	—	—	33.5	1.9	42.5	166.25	1.25	8.75
Average	3.5	13.1	15.6	69.4	3.5	37.5	—	4.0	35.6	2.8	22.6	122.9	2.2	6.7

Average														
1	—	13.0	—	66.6	—	33.3	—	—	44.5	8.3	—	36,7	—	3.3
2	1.9	14.1	52.9	47.1	1.9	45.1	—	5.5	27.5	4.1	3.8	57.2	2.25	2.8
3	6.8	11.7	26.2	73.8	7.6	31.9	—	8.6	30.4	3.0	35.5	122.2	3.2	7.9
4	7.8	10.2	50.9	33.7	2.8	30.1	15.4	—	40.9	8.2	21.0	160.9	6.6	3.4
Total	4.6	12.3	35.0	55.8	3.7	35.3	3.6	4.3	34.4	5.4	17.6	99.6	3.2	4.7

TABLE 4.3 : **Yield per ha and per tree in coconut**

Size Group	*Nuts/ha*	*Nuts/tree*	*Nut/yielding tree*	*% of yielding trees*
Category I				
1	3736	21.0	27.0	77.8
2	3451	19.4	26.0	74.6
3	4572	25.7	34.0	75.6
4	3380	19.0	23.0	81.5
Average	3867	21.7	28.3	77.1
Category II				
1	6939	39.0	47.0	83.0
2	5764	32.4	39.5	82.0
3	5231	29.4	37.0	79.5
4	5159	29.0	39.0	74.4
Average	5629	31.6	30.8	80.0
Average				
1	4804	27.0	33.7	79.5
2	4376	24.6	31.4	77.6
3	4792	26.9	35.0	76.9
4	4064	22.8	29.3	68.8
Total	45.08	25.3	32.5	78,0

TABLE 4.4 : Input quantities of coconut per ha.

Size Group	*Family labour (mandays)*	*Hired labour (mandays)*	*Organic manure (kgs)*	*Fertiliser (kgs)*
Category I				
1	44	—	1600	15.0
2	10	35	800	41.5
3	14	45	800	75.0
4	—	60	800	50.0
Average	14.9	38.1	937	50.4
Category II				
1	40	20	1200	—
2	34	30	800	45.4
3	40	24	1080	51.3
4	10	50	800	30.0
Average	30.7	31.7	944	36.5
Average				
1	42.6	6.6	1467	10.0
2	19.6	33	800	43.1
3	22.6	38	893	67.0
4	3.8	56.2	800	42.3
Total	20.6	35.7	940	45.3

TABLE 4.5: **Cost of cultivation of coconut (Rs/ha)**

Size Group	*Family labour*	*Manure*	*Fertiliser*	*Electricity*	*Diesel*	*Total cost*	*Total income*	*Subsidiary income*	*Benefit cost Ratio*
Category I									
1	660	180	38	—	—	1926	7472	—	2.87
2	670	100	104	23	5	1984	6903	1000	2.47
3	885	100	188	28	2	2646	9147	—	2.45
4	900	100	125	20	1	2520	6768	—	1.68
Average	795	113.7	126.2	20.1	2.2	2323	7739	257	2.34
Category II									
1	900	150	—	—	—	2310	13877	—	5.0
2	960	100	114	23	2	2636	11529	500	3.37
3	960	130	129	28	—	2742	10461	400	2.82
4	900	100	75	20	—	2408	10319	200	3.28
Average	936	116.5	91.7	20.3	0.6	2581	11258	320	3.37

Average									
1	738	170	25.3	—	—	2054	9607	—	3.67
2	786	100	108	23	3.8	2245	8753	800	2.89
3	910	110	168	28	1.3	2678	9585	133	2.57
4	900	100	105.8	20.3	0.6	2477	8134	77	2.28
Total	845	114.7	113.6	20.2	1.6	2417	9019	280	2.71

TABLE 4.6 : **Nature of coconut disposal**

Size Group	*(Percentage disposal into)* Tender coconut	Coconut for eating	Sales	Copra
Category I				
1	2.13	38.9	39.8	—
2	4.7	40.1	42.8	12.4
3	2.0	14.2	72.5	11.3
4	1.8	19.0	65.0	14.2
Average	6.0	26.2	57.5	10.3
Category II				
1	2.5	23.7	48.5	25.3
2	4.8	13.1	71.1	11.0
3	3.1	15.3	64.5	17.1
4	2.0	13.7	68.8	15.5
Average	3.2	16.5	65.2	15.1

5

Arecanut

Arecanut palm, *Areca catechu L* is cultivated primarily for its kernel obtained from the fruit. The edible endosperm of areca fruit is used a masticatory in india and South-east Asia either alone or as a constituent of the quid along with betel leaves lime and tobacco. It can be chewed in tender, ripe or processed form. It is essentially a tropical crop. A well distributed rainfall of about 250 cm is ideal for the crop but it is found in low rainfall regions also. For good performance of arecanut crop, uniform distribution of rainfall, a well drained soil and facilities for irrigation during the dry period are essential. During summer the high temperature is to be counteracted by profuse irrigation. Though it can be grown upto 900 m from MSL lower attitudes are preferable. Nuts produced at higher attitude are found unfit for processing into desirable quality because of reduced size, excessive shrinkage and cracking. Arecanut palms are grown in different soils such as literite, red and sandy loam, alluvial and deltaic. The soil should be well drained and capable of retaining optimum moisture. It is preferable that the soil is slightly acidic.

Arecanut palms are highly prone to sun-scorch and wind. Therefore proper alignment of the garden to minimise these hazards is essential. Normally about 1250 palms per hectare of garden can be accommodated. The seedlings are to be protected from the sun in the early stages by providing artificial shade or by growing intercrops like banana. The normal planting seasons is May-June but in clayey soils and water-logged areas planting is undertaken during August-September. Because of heavy rainfall in most of the areca growing areas, the soils might be severely leached and eroded. Hence the requirement of organic and synthetic manure in large quantity becomes essential. According to experimental results fertilisers to supply 100 g N, 40 g P_2O_5 and 104 g of K_2O and organic manure like cattle dung, compost and green manure at the rate of 12 kg per bearing palm are essential. About 50.0 per cent of the manures is to be applied in September and the rest in February. Apart from manures, lime may be applied as the soils are acidic in reaction. Growing of cover crops may prevent soil erosion and weed growth besides augmenting organic manure supply to the soil. Arecanut can be pure or mixed crop depending upon the local conditions. The crops which could be cultivated without affecting the prospects of main crop are banana, betalvine, cocoa, ginger, pepper, etc. Crops like banana, yam, tapioca, pineapple, arrow-root, fodder crops and fruit trees like mango and jack in the areca fields along with coconut inside (in case of paddy displacement) or as a border crop are usually mixed.

Depending upon the consumption requirement, arecanut is harvested at tender stage or when the nuts are fully riped. Fully ripe arecanut (challi) are harvested during November to February and sun-dried for about 40 days by evenly spreading them in single layer on level ground or cement floor. Uniform dried arecanuts are turned over at weekly intervals. The dried arecanut or the dehusked percentage recovery of dried husked nuts varies from 22 to 25. Tender nut is processed in farms in South India. It is also used for converting into scented supari. Nuts of 6 to 7 months maturity are harvested for tender nut processing. There are several types of processing About 10 to 12 kg of dried product is obtained after processing 100 kg of unhusked green nuts with proper management. Arecanut palms start yielding from the sixth year. Stabilised yield is obtained after 10 years and is maintained till about the 40th year. The yield declines then-after and becomes uneconomical by the 60th year. It is usual to find areca seedlings planted as a stand for replacement of the old trees. Recent varieties include improved cultivars which are capable of yielding double the local level.

Yellow leaf disease has been causing much damage to arecanut crop in Kerala and parts of Karnataka in recent times.

The disease is characterised by yellowing of leaves which progresses along with the margin of leaflets leaving their mid-rib green. As a result, reduction in the size of leaves and nuts, tapering of the stem and mature nutfall occurs, colour of the kernel inside the nut turns brown and it becomes unsuitable for consumption. The palm dies gradually. The etiology of the disease is not clearly understood and hence there is no effective control measure for it. Water, logging, acidity, nutritional disorders, etc, may be responsible for it. Inter and mixed cropping practices in the disease-affected garden can provide additional income to the farmer. It is estimated that about 35-40.0 per cent of areca garden in Kerala has been affected by this disease (especially in the southern parts of the state).

Koleroga (fruit-rot) is another disease prevalent in high rainfall regions. The disease assumes intensity during south-west monsoon causing heavy damage to the crop. In the initial, water socked patches appear where the colix of the affected nuts and the nuts become darker in colour, finally, the nuts drop down and the fallen nuts are covered by the felt of myclial growth. Bud-rot, food-rot, stem breaking, inflorescence die-back, stem bleeding, anaberoga are other diseases which affect areca yield and cause damages in varying degrees. Control measures for the above diseases include spraying the bunches with 1 per cent bodeaux mixture in intervals.

The important pest of arecanut is spindle bug. It lives in the inner-most leaf axil and suck the sap from the tender leaves and spindle. Severely attacked spindles neither attain normal size nor open completely. Red and colyx mites attack seedlings during summer; inflorescence caterpiller and root grut are the other the pests. To prevent pests, proper chemical control measures have been suggested in varying degrees.

Trends in area and Productivity

India is the major producer of arecanut. India produces about 88.0 per cent of world arecanut. Other countries where arecanut is cultivated as a minor crop are Bangladesh, Indonesia and some of the Pacific Ocean Islands. Use for arecanut other than chewing is negligible. Its exports prospects are also very much limited. About 90.0 per cent of the area and 95.0 per cent of production are concentrated in Kerala, Karnataka and Assam. Arecanut area in the country increased from 1.17 lakh ha in 1961-62 to 1.67 lakh ha 1970-71, to 1.85 lakh in 1980-81 and 1.87 lakh ha in 1983-84. It may be seen that 1970-71 onwards increase in area has been marginal. Intensive cultivation has resulted in increased productivity. The yield increased from 813 kg/ha in 1961-62 to 843 kg/ha in 1970-71 and 1037 kg/ha 1980-81. The incidence of yellow leaf disease in Kerala has considerably reduced the yield rate in that state as a result of which the national average declined to 994 kg/ha in 1983-84. During this year the yield rate except in Karnataka has not been encouraging in other states. Karnataka has been showing a satisfactory performance in arecanut cultivation. Area under arecanut increased from 0.41 lakh ha 1970-71 to 0.45 lakh ha in 1980-81 and to 0.57 lakh ha in 1985-86. Unlike the nation, average yield level in Karnataka has been quite significant. The yield level increased from 1229 kg/ha 1970-71 to 1459 in 1980-81 and 1462 kg/ha in 1983-84 but slightly declined to 1450 kg/ha in 1985-86. Kerala with an area 0.6 lakh ha and yield level of 685 kg/ha and Assam with an area of 0.5 lakh ha and the yield rate of 954 kg/ha were the other two arecanut states. During 1983-84, total area under arecanut in the country was 0.19 mil ha and Kerala, (0.06 mil ha), Karnataka (0.06 mil ha and Assam 0.05 mil ha) accounted for over 89.4 per cent of the area.

Age and Type of Gardens

Table 5.1 shows talukwise arecanut area in D.K. It shows that there is a marginal increase in area in most of the taluks. Sullia, Puttur, Bantwal and Belthangady are the major arecanut areas in the district (8 to 20 per cent of gross cropped area). In arecanut also age classification has been attempted as under less than 7 years, 7-15 years, 15-30 years and above 30 years. In the sample area all the farmers were growing the local varieties of arecanut. Recently some have started growing the improved VTL cultivars like Mangala which yield more than double the locals. As the proportion was fouud to be very low, for the improved strains, the entire sample area has been treated as falling under local varieties. In arecanut pure and mixed type of gardens were existing. The proportion of pure type of gardens was 71.5 per cent in the first category and 81.8 per cent in the second. In the first group of first category the entire gorden was of mixed type. In such cases the trees mixed were banana, mango, jack, coconut, pepper and others. If the particular mixed crop did not yield over 10.0 per cent of the total agriculture income the same has been omitted from calculation under pure type. In the first category the fourth group has no mixed type of arecanut gardens. The proportion of mixed type of garden was very low in the II group and also II group of second category (Table 5.2). The IV group in the first category and I group in the second category had only pure type of gardens. The proportion was slightly lower in the first in the second category. The different age-group classification presents that the proportion was 43.8 per cent and second and third groups showed mixed type of garden. The 30+age-group shared 1.5 per cent in the first and 6.7 per cent in the second category. According to Table 5.3, the proportion of yielding trees was 77.6 per cent in the first and 74.0 in the second category. The proportion was evenly spread out except in the last group of first category and second group of second category where it was about 64-65.0 per cent. The proportion of non-yielding trees (23.6 per cent) was under non-bearing stage. Because even the non-bearing trees were included in the areas, yield of arecanut has been considered as per total member of trees. The yield per yielding trees was 1.8 kg processed nut in first and 2.17 kg in the second category. The yield was low in the first three group of the first category. For the entire sample the first and third groups reported lower yield levels. In terms of total number of trees the yield per tree was 1.34 kg in the first and 1.59 kg in the second category. The first and third groups for both categories combined had a lower yield level than the sample average of 1.42 kg. On this basis the per ha productivity worked out to 1683 kg in the first and 1990 kg in the second category and for both combined 1974 kg per ha. All the groups except the fourth in the first and the second in the second category had lower productivity. For the combined sample, groups I and III had lower yield rates. If we consider only the bearing palms, the average productivity would be 2425 kg per hectare. For the Dakshina Kannada district the average yield during 1983-84 was 1802 kg as against 1450 kg for the state, In terms of raw nut, the productivity was higher by 4.69 times than the processed nuts in the district and 4.8 times in the state indicating a better ratio for processed nut in Dakshina Kannada over the stage average.

Economics of Arecanut Cultivation

Under proper management, arecanut crop start yielding from the sixth year. Stabilised yield is obtained after ten years and is maintained till about the 40th year. The yield declines thereafter and becomes uneconomical by the 60th year. Under proper management conditions the yield of arecanut varies from 1500 to 3500 kg/ha. The price Mangalore supari (fully matured sundried husked arecanut) which was Rs. 2600/t in 1956 increased to Rs. 4600 in 1960, to Rs 6600 in 1970 and to Rs. 15090 in 1980. In the early eighties, the price shot up and rose upto Rs. 30,000/t. When India was partitioned, a sizable arecanut area went to Bangladesh. Indian arecanut used to be smuggled to countries like Pakistan and recent extended vigilence in the borders has reduced the demand for smuggling. About 70.0 per cent of arecanut is sold in the processed form. Boiled (tender and semi ripe fruits) and unboiled (ripe nut of 10 to 11 months old) are the two different types. About 72.0 per cent of the farmers prefer sales of fruit to traders in the village or primary market. About 6.0 per cent of the farmers effected the sales through commission agents, 5.0 per cent through regulated markets and 17.0 per cent through CAMPCO (Central Arecanut and Cocoa Marketing and Processing Cooperative Society). It has been estimated that the farmers receive about 55.0 per cent of the consumers price. Arecanut which is more than one year old loses its protein, sugar and total fat contents, but shows increase in the alkaloid content. The one year old nuts demand a better price than the fresh ones. During the reference period, the farmers who had old arecanut expecting a higher price suffered due to crash in prices. The price increase was sought by the efforts of CAMPCO and it is only r.cently that there has been an increasing trend in the prices.

In the cost of cultivation, labour cost accounted for the bulk of the total cost in arecanut cultivation. It was 54.2, 50.1 and 52.1 per cent in small, medium and large gardens respectively. Material cost was the next important item sharing 33.1, 33.2 and 29.6 per cent respectively in the three size groups. Home consumption of arecanut was negligible and accounted for less than 1.0 per cent of total production.

According to Table 5.4, for the sample on an average, per ha labour requirement was about 237 mandays. The IV size group in the first category had reported a larger labour input, when compared to other groups. In the second category also the per ha labour requirement was around 333 mandays. The proportion of hired labour was about 60.8 and 62.5 per cent in the first and second categories respectively.

Organic manure application was in the order of 1620 kg/ha the first category (high dosage in the earlier groups) and 1666 kg/ha in the second (the first and third groups leading others). Fertiliser application was substantial in the last group (24/ kg/ha) of first category and in the second and third groups (353 and 430 kg/ha), of second category. The average fertiliser consumption has been 158 kg/ha and 276 kg/ha respectively. The first group of the second category did not consume any fertiliser and first two groups in the first category consumed a lower dosage (Table 5.5).

Arecanut palms in all the size groups were protected against the attack of pest and diseases. It has been observed to be uniform throughout the groups and shared about Rs. 1000 per ha. The cost also included electricity, diesel and other cost including interest, rental value, etc. On this basis total cost of cultivation in the average farm came to about Rs. 9159/ha and Rs. 9668/ha in the two categories. The first group in both categories and the second in the first category had a lower expenditure level. The proportion of labour cost was 56.9

and 51.5 per cent respectively in the two categories. Between the groups, the variation was not much—54.0 per cent in the third and 60.9 per cent in the first group of first category and 49.8 per cent in the third to 58.3 per cent in the first group of second category. The first group in both the categories indicated a larger share for labour when compared to others.

In terms of net BCR, the first category had a lower ratio (2 7) when compared to the second (3.11). The ratio was 2.84 for both combined. The II (2.92) and IV (3.67) groups performed better than the sample average while the I (2.15) and III (2.43) groups performed poor. In the first category, the IV group and the II group in the second category had very good performance.

Besides the main crop income the farmers reported income from subsidiary crops which was Rs. 2376 and Rs. 1120 per ha in the two categories respectively. It may be pointed out that during the reference period, the price of arecanut was lower to preceding or succeeding years and this had resulted in relatively a lower BCR.

TABLE 5.1 : **Arecanut area in different taluks of D.K. (in ha)**

Taluk	*1983-84*	*1985-86*	*Proportion to Gross cropped area %*
Bantwal	3925	4035	10.7
Belthangady	2500	3063	7.9
Karkal	1086	1132	2.7
Kundapur	869	959	2.1
Mangalore	673	688	2.1
Puttur	3565	3586	13.3
Sullia	4090	4045	19.7
Udupi	421	433	0.8
Total	17.129	17,941	6.6

TABLE 5.2 : **Age and variety in arecanut (%)**

Size Group	*Less than 7-years*		*7-15 years*		*15-30 years*		*Above 30 years*		*Total no. of trees*	
	Pure	*Mixed*	*Pure*	*Mixed*	*Pure*	*Mixed*	*Mixed*	*Pure*	*Pure*	*Mixed*
Category I										
1	—	63.6	—	18.2	—	18.2	—	—	—	275.0
2	14.5	—	47.6	17.2	20.7	—	—	—	300	62.5
3	21.4	9.8	14.3	22.3	20.5	7.1	—	4.6	525	408.4
4	32.0	—	50.0	—	18.0	—	—	—	1100	—
Average	18.4	14.2	28.6	15.2	16.5	5.6	—	1.5	508.6	203.0
Category II										
1	22.2	—	33.3	—	—	—	44.5	—	900.0	—
2	19.1	3.6	16.4	20.9	36.4	3.6	—	—	987.5	387.5
3	21.5	—	38.2	—	33.3	7.0	—	—	1675.0	125.0
4	22.2	—	28.8	9.6	24.1	15.4	—	—	1950	650.0
Average	21.0	1.1	28.6	8.6	26.9	7.0	6.7	—	1421.3	316.25

Average										
1	7.4	42.4	11.1	12.1	=	12.1	14.9	—	300.0	183.3
2	16.3	1.4	35,1	18.6	27.0	1.4	—	—	575.0	192.5
3	21.4	6.5	22.3	14.8	24.8	7.1	—	3.1	908.0	313.9
4	28.3	—	41.8	3.7	20.3	5.9	—	—	1426.9	250.0
Total	19.3	9.5	28.6	12.8	20.3	6.1	2.4	1.0	840.5	244.2

TABLE 5.3 : **Yield per ha and per tree in arecanut**

Size Group	*Kg/ha*	*Kg/tree*	*Kg/Yielding tree*	*% of yielding trees*
Category I				
1	1000	0.8	1.0	80.0
2	1463	1.17	1.31	89.3
3	1563	1.25	1.4	76.2
4	2625	2.1	3.25	64.6
Average	1683	1.34	1.81	77.6
Category II				
1	1875	1.5	1.9	78.9
2	2275	1.82	2.87	63.4
3	1838	1.47	1.82	80.8
4	1900	1.52	1.95	77-9
Average	1990	1.59	2.17	74.6
Average				
1	1291.7	1.03	1.3	79.6
2	1787.8	1.43	1.93	78.9
3	1654.7	1.32	1.7	77.7
4	2346.2	1.87	2.75	69.7
Total	1795	1.43	1.94	76.4

TABLE 5.4 : **Input quantities of arecanut per ha**

Size Group	*Family labour (mandays)*	*Hired labour (mandays)*	*Organic manure (kg)*	*Fertiliser (kg)*
Category I				
1	247	82	1960	64
2	81	238	1960	90
3	148	191	1480	198
4	99	309	1240	247
Average	136	211	1630	158
Category II				
1	161	165	2640	—
2	91	247	1520	353
3	178	154	1880	430
4	79	250	1000	165
Average	125	208	1666	276
Average				
1	218	109	2187	43
2	85	241	1784	195
3	158	179	1613	275
4	91	286	1148	215
Total	132	210	1644	201

TABLE 5.5 : Cost of Cultivation of Arecanut (Rs/ha)

Size Group	*Family labour*	*Hired labour*	*Organic manure*	*Ferti-liser*	*Electri-city*	*PP Chemicals*	*Diesel*	*Subsidiary income*	*Total cost*	*Income*	*Benefit cost ratio*
Category I											
1	3704	1230	245	160	—	1000	—	3295	8100	20000	1.47
2	1215	3570	245	225	62.0	1000	10	2246	8612	29260	2.39
3	2220	2872	185	495	68.6	1000	8	2363	9433	31260	2.31
4	1480	4635	155	618	69.2	1000	4	1853	10157	52500	4.16
Average	2047	3173	204	396	55.3	1000	6.2	2376	9159	33670	2.70
Category II											
1	2410	2475	330	—	—	1000	—	—	8373	37500	3.37
2	1360	3705	190	883	62.0	1000	10	1841	9830	45500	3.62
3	2671	2310	235	1075	68.6	1000	8	1594	10004	36760	2.67
4	180	3750	125	824	69.2	1000	4	357	9847	38000	2.85
Average	1866	3113	208	793	56.5	1000	6.4	1120	9668	39803	3.11

Average											
1	3273	1645	273	107	—	1000	—	2197	8191	25833	2.15
2	1273	3624	223	488	62	1000	10	2084	9099	35756	2.92
3	2370	2685	202	688	68.6	1000	8	2108	9623	33093	2.43
4	1365	4295	143	297	69.2	1000	4	1278	10037	46923	3.67
Total	1981	3151	205	540	55.7	1000	6.3	1119	9344	35900	2.87

6

Paddy

Paddy is an important crop sharing more than 50.0 per cent of the gross cropped area in the district. The proportion under high yield varieties has been about 30.0 per cent in kharif, 54.0 in rabi, 75.0 per cent in summer and 40.0 per cent on average. This is considerably lower than the state average of 56.0 per cent. The yield of HYVs has been 3664 kg/ha in kharif (3604 kg for the state), 3346 kg/ha in rabi (similarly to the average) and 3312 kg/ha in summer (as against 3802 kg in the state). For all the seasons combined, the average yield of local varieties has been 1993 kg/ha and that of HYVs 3336 kg/ha giving an overall paddy yield of 2510 kg/ha. The corresponding figures for the state are 2128 kg, 3450 kg and 2866 kg/ha respectively. The lower proportion of irrigation and heavy rainfall regime and other local factors are responsible for the below average performance. During rainy season too much of water-logging affects the period and quality of production not only in paddy but in other crops also.

In the district the proportion of paddy area in kharif is about two-thirds of the annual area. The remaining area is distributed between the rabi and summer in the ratio of 86 : 14. Table 6.1 describes paddy area in 1983-84 and 1985-86 in respect of different taluks. Except Sullia, all the taluks have substantial area under paddy, the bulk being in kharif. Total paddy area in the district increased from 0.15 mil ha in 1983-84 to 0.16 mil ha in 1985-86.

Paddy cultivation in the sample farms

Paddy was cultivated in about 63.0 per cent of the sample farms. That is, some farmers did not grow paddy but concentrated on arecanut. As a matter of fact, uneconomic paddy and vegetable cultivation leads to conversion of paddy lands into gardens. If coconut is found in some of the areca gardens it is due to the fact that it was a converted paddy land and since coconut was a border crop in small paddy plots, the same becomes inside the big areca garden.

In the first category which cultivated paddy, paddy area constituted about 31.6 per cent of total crop area. The proportion was higher in the second group (36.5 per cent); in all others it was average or less. Paddy is a labour intensive crop requiring large quantities of human and bullock labour, manure and fertilisers and water. The labour requirement per ha of paddy was 164 mandays and 58 bullock days in the first, 132 mandays and 36 bullock days in the second, 171 mandays and 45 bullock days in the third and 119 mandays and 39 bullock days in the fourth size groups. On average, it was 148 mandays and 44 bullock days for the entire sample. Of the total human labour, about 70.0 per cent was hired, the proportion increasing for the larger size groups. In organic manure and fertiliser consumption, larger groups usually consumed a larger quantity (407 and 308 kg in III and IV groups) as against 247 kg in the first two groups. (Table 6.2).

The total cost of cultivation in the first group was Rs. 5279 per ha of which 47.0 per cent was shared by human labour and bullock labour; the share was 67.5 per cent in the second, 66.6 per cent in the third and 62.8 per cent in the fourth size groups. The cost of cultivation was lower in the second group at Rs. 4265, (due to lower labour input) increasing to Rs. 5540 in the third and declining to Rs. 4395 in the last group. Besides labour, the other items of expenditure include manure and fertiliser (about 21.0 per cent of total cost), expenses on diesel and electricity, plant protection chemicals, inputed cost of rental value, interest, etc. (Table 6.3).

The average yield of paddy has been 5058 kg per ha and in the different groups 3624 kg, 4876 kg, 5726 kg and 5336 kg respectively. Higher fertiliser consumption in the larger size groups could have been responsible for higher productivity. At the existing market prices of paddy, total income in the different size groups was Rs. 9060, Rs. 12190, Rs. 14316 and Rs. 13342 respectively and on average Rs. 12646 per ha. In terms of net benefit cost ratio the average performance was lower in paddy (1.70) when compared to coconut or arecanut. The ratio was only 0.72 in the first group and in the third group it was slightly below the sample average at 1.58. The second (1.86) and fourth (2.04) size groups showed a favourable rate of performance.

About 25.0 per cent of the farmers grew paddy in the summer but the net benefit cost ratio was about one only (the average cost of production was 5578 and income Rs. 11118). During summer, the total labour requirements increased to 69.0 per cent of total cost. The per ha expenditure on manure and fertiliser was also high but the average productivity was about 12.0 per cent lower than in kharif. Table 6.4 describes cost particulars for kharif and summer paddy crops in the case of selected samples.

Paddy was cultivated in the second season only in the case of about 25.0 per cent of the farmers. In other cases, vegetables or pulses mainly for own consumption were raised. The average income in this respect was Rs. 1370, it being Rs. 2295, Rs. 1089, Rs. 1395 and Rs. 954 respectively in the different groups.

Totality of the Samples

An attempt is made in this section to consolidate different crops and arrive at farm economics. The farmers cultivating paddy, coconut and arecanut and only arecanut and coconut along with minor crops have to be viewed in totality to so as to obtain the total cost, income and net benefit cost ratio of the farm.

The average area in the different groups was 0.71, 1.35, 2.10 and 2.62 ha in the first category and 0.79, 1.37, 1.63 and 2.19 ha in the second category respectively. Table 6.5 presents cost of cultivation per ha of coconut, arecanut and paddy and total cost for all the crops combined. The average cost of cultivation per ha in different groups was Rs. 3855, Rs. 7119, Rs. 13145 and Rs. 18191 in the first category and Rs. 5099, Rs. 10378, Rs. 11877 and Rs. 15985 in the second category respectively. The average cost in the case of paddy farmers worked out to Rs. 10896 and in the case of non-paddy farmers it was slightly higher at Rs. 11352. The income figures also showed higher figures in respect of the non-paddy farmers, when income from coconut arecanut and paddy along with subsidiary crops were consolidated (Table 6.6). The average income was Rs. 10268, Rs. 24896, Rs. 43049 in the first and Rs. 23719, Rs. 49673, Rs. 45458 and Rs. 63123 in the second category respectively. The lower groups reported lower income levels possibly because of a lower share of arecanut and coconut. The non-paddy farmers and also the big paddy farmers were largely commercial and reported very high income levels. The benefit-cost ratio was 1.66, 2.48, 2.54 and 3.59 in the first category and was higher in the second category at 3.65, 3.78, 2.82 and 2.94 in the different groups respectively. This shows that the average net benefit-cost ratio was higher in the case of non-paddy farmers (3.26) over paddy farmers (2.54).

Table 6.7 presents the benefit-cost ratio for different groups in the sample when both paddy and non-paddy cases were combined. The average area in the different groups was 0.74, 1.36, 1.95 and 2.57 ha with a combined benefit-cost ratio of 2.36, 3.00, 2.60 and 3.38 respectively. The average ratio was 2.77 for all the cases indicating a viable farming system in the area.

Effect of Rainfall

A major factor in the stability of production of the crops is impact of monsoon, others being under-planting or over-planting, inadequate fertilisation, low fertile soil, less appropriate development in genetics and incidence of pest and diseases. For example, in some years, the yield rate of coconut is as high as 10,000 nuts/ha, the maximum any nation can produce (as in Indonesia) but in many years the yield level falls to as low as 1500-2000 nuts/ha. The distribution of the preceding year rainfall may have a contributory factor. In drought years or in years where rainfall is too much concentrated, the succeeding year yield level would be low for the rainfed crops. The yield rate may not be influenced negatively by normal rainfall distribution. The main harvesting season for coconut is between January to July. If rainfall is normal especially in the preceding monsoon season, coconut productivity may not be affected. Where minimum rainfall is reported over years (as in Hassan) or where bulk of the crop is irrigated (as in Mysore) coconut productivity has remained more or less stable except in drought years. In Dakshina Kannada also, poor rainfall years are marked with slight variation in coconut production. In rainfed regions like Tumkur, poor rainfall may be damaging. Between 1975-76 and 1980-81 a stable rainfall regime has resulted in stability in productivity. After 1981-82, increase in the probability of monsoon failure has resulted in notable variations in productivity. Whereas in Kerala where the wilt disease was prevalent, the productivity of coconut has been affected much. Traditionally when paddy fields are converted into gardens arecanut plays a leading role because of its heavy water duty nature unlike coconut. Nonetheless, both arecanut and coconut area have increased over years but the proportion of increase is very high in the case of arecanut in the coastal districts than others. In Dakshina Kannada for example, the increase in coconut area is by 35.0 per cent in 1980-81 over 1955-56 and in Uttar Kannada by a mere 6.0 per cent. The increase in arecanut area is by over 14.0 and 18.0 per cent respectively in the two districts over the above period. The compound growth rates of area, production and yield of arecanut in the state have been 2.6, 6.0 per cent and 3.4 per cent respectively per annum and in coconut 2.6, 3.5 and 0.9 per cent respectively. Though coconut and arecanut have increased at a similar rate the productivity gains have been favourable to arecanut. The incidence of pest and diseases, declining soil fertility and competition from crops like arecanut and rubber in the specified regions may be responsible for the poor performance of coconut in these districts.

TABLE 6.1 : **Paddy area in D.K. district (ha)**

Taluk	*1983-84*				*1985-86*			
	Kharif	*Rabi*	*Summer*	*Total*	*Kharif*	*Rabi*	*Summer*	*Total*
Bantwal	11269	6631	1765	19665	11254	7140	1458	19852
Belthangady	9207	5425	475	15107	10785	6615	427	17827
Karkal	17195	7219	1274	25668	17054	6800	1170	25024
Kundapur	19453	3337	311	23101	19880	3544	261	23685
Mangalore	10195	5293	1610	17098	10275	7760	2366	20401
Puttur	10969	4607	353	15929	7043	4441	245	11729
Sullia	2693	1920	67	4680	2772	1894	87	4755
Udupi	23634	5686	1529	30849	24760	6391	1353	32504
District	101,602	40,198	7,384	149,184	103,823	44,585	7,367	155,775

TABLE 6.2 : **Input qoantities of paddy (kg/ha)**

Category	*Family labour (Mandays)*	*Hired labour/ (Mandays)*	*Bullock labour (Days)*	*Organic manure*	*Fertiliser*
1	106	58	58	1960	247
2	29	103	36	1640	247
3	29	142	45	1760	407
4	40	79	39	1800	378
Average	44.7	103.2	43.5	1773	331.8

TABLE 6.3 : **Cost of Cultivation of Paddy (Rs/ha)**

Category	*Family labour*	*Hired labour*	*Bullock labour*	*Organic manure*	*Fertiliser*	*Electricity*	*Oil*	*P.P.C.*	*Total cost*	*Income*	*Benefit cost ratio*
1	1590	870	1450	245	617	—	—	50	5279	9060	0.72
2	435	1545	900	205	617	45.0	35	50	4265	12190	1.86
3	435	2130	1125	220	1018	42.3	17	50	5540	14318	1.58
4	600	1185	975	200	944	34.0	9	50	4395	13342	2.04
Average	671	1548	1088	218	830	30	18	50	4905	12646	1.70

TABLE 4.4 : **Cost of Cultivation of Paddy-Kharif and Summer (Rs/ha)**

Item		*Kharif*	*Summer*
Labour :	family	762	1026
	hired	1529	1596
	animal	960	1425
Manure :	organic	189	266
	synthetic	807	860
Others :	Electricty	30	30
	Oil	18	25
	Plant Protection	50	70
Total cost		4905	2578
Total income		12,646	11,118
Benefit-cost ratio		1.70	1.00

TABLE 6.5 : **Cost of cultivation per farm (all crops) (Rs)**

Category I Group	*Area (ha)*	*Coconut cost/ha*	*Arecanut cost/ha*	*Paddy cost/ha*	*Total cost/ha*
1	0.71	1926	8100	5279	3855
2	1.35	1984	8612	4265	7119
3	2.10	2646	9433	5540	13145
4	2.82	2520	10157	4395	18191
Average	1.85	2323	9159	4905	10896
1	0.79	2310	8373	—	5099
2	1.37	2636	9830	—	10378
3	1.63	2742	10004	—	11877
4	2.19	2408	9847	—	15986
Average	1.57	2581	9568	—	11352

TABLE 6.6 : **Total Income from all Crops (Rs/ha)**

	Coconut		*Arecanut*		*Paddy*		*Income/ha*
	Main	*Sub*	*Main*	*Sub*	*Main*	*Sub*	
Category I							
1	7472	—	20000	3295	9060	2295	10268
2	6903	1000	29260	2246	12190	1089	24806
3	9147	—	31260	2363	14316	1395	43049
4	6768	—	52500	1853	13342	954	83592
Average	7739	257	33670	2376	12646	1370	38622
Category II							
1	13877	—	37500	—	—	—	23719
2	11529	500	45500	1841	—	—	49672
3	10461	400	36760	1594	—	—	45458
4	10319	200	38000	357	—	—	63123
Average	11,258	320	39,803	1120	—	—	48,403

TABLE 6.7 : **Benefit cost ratio for all crops**

Group	*Area (ha)*	*Paddy BCR*	*Arecanut BCR*	*Coconut BCR*	*Combined BCR*
1	0.74	0.72	2.15	3.67	2.36
2	1.36	1.86	2.92	2.80	3.00
3	1.95	1.58	2.43	2.57	2.60
4	2.57	2.04	3.67	2.28	3.38
Average	1.75	1.70	2.84	2.71	2.77

7

Cardamom, Pepper and Rubber

Cardamom

Small cardamom (*elettaria cardamom*) is the queen of spices and India is the largest producer and exporter of this commodity. It accounts for about 70.0 per cent of production and 60.0 per cent of world trade. It is cultivated as pure or subsidiary to coffee or arecanut plantations. It is also found as natural undergrowth and the some is abandoned after a few years when productivity becomes very low. The crop is sensitive to changes in growing conditions and hence world production varies substantially over years.

Kerala is the largest producer of cardamom and its share in area increased from 50.8 per cent in 1963-64 to 64.3 per cent in 1980-81. But the yield rate has been always lower than the national average (varying from 25 kg/ha in 1970-71 to 10 kg/ha in 1980-81). Karnataka is the next improtant cardamom growing state with 30.7 per cent share. Its share in production declined from 50.0 per cent in 1963-64 (and 1975-76) to 33.0 per cent 1980-81. Tamil Nadu has the highest yield rate at 95-kg/ha (over 60.0 per cent more than Kerala and Karnataka).

The total cardamom area is distributed among more than 18,500 holdings, majority being of less than 4 ha. Export of cardamom was 54.0 per cent of production in 1970-71 and it increased to 71.0 per cent during 1979-80.

In Karnataka, Kodagu, Hassan, Chikmagalur, Dakshina Kannada and Uttara Kannada are important cardamon growing areas. Over years there has been a study increase in area in the case of Kodagu (from 7206 ha in 1960-61 to 9409 ha in 1965-66, 11301 ha in 1970-71 and 24058 ha in 1975-76) but the area declined after 1980-81 (14840 ha). In other districts there has been an area decline (from 4777 to 4340 ha in Chikmagalur, 8434 to 7072 ha in Hassan, 351 to 228 ha in Uttara Kannada and 369 to 357 ha in Dakshina Kannada in 1980-81 over 1960-61). The state area increased from 21218 to 37318 ha in 1975-76 and declined to 26863 ha in 1980-81. The yield rate declined from 71 kg in 1960-61 to 59 kg in 1980-81, the decline was common for all the districts. In Dakshina Kannada for example, it declined from 98 to 45 kg/ha during the period. Recently in Kundapur taluk of the district, new cardamom plantations have been started as a result of which cardamon area in the district increased to 1373 ha in 1983-84.

The major problem in cardamom is one of declining yield level. Declining productivity and increase in cost of production naturally limit competitiveness and viability. Cardamom production declined from 4800 mt. in 1980-81 to 3500 mt. in 1984-85. The quantity exported also declined to 67.0 per cent from 71.0 per cent in 1977-78. Though there has been competition from countries like Guatemala, Indian cardamom continues to occupy an important place due to its superior quality. Genetic manipulations, fixing green colour in varieties grown in areas like Karnataka, product diversification into cardamom oil, etc, and control of the mosaic (katte) disease may strengthen the yield level and subsequently improve the competitive nature and export performance. Consolidation of small holdings into mechanisable ones would also be a right direction toward the above objectives.

Pepper

Pepper (*piper nigrum L*), the king of spices is a most important and earliest known spices. It had played a historic role in enabling Europeans to discover India in the wake of pepper shortage in the middle ages. Pepper had originated in the south western India and is closely allied to betalvine. It is a permanent climbing plant and cultivated as pure or mixed with coffee, orange or areca palms. It requires a fully tropical climate with a minimum rainfall of about 150 cm. and a humid warm climate. There are more than 75 varieties under cultivation and recently an hybrid variety (Ranniyar 1) has been gaining importance. Pepper is marketed in the form of black pepper (in its natural form), white pepper (deskinned), pepper powder, deoresil (prepared by solvent extraction of ground pepper), pepper oil, green pepper (bottled) and dehydrated green pepper. Pepper flowers in May-June and the crop matures in about 6-8 months after flowering. During harvesting the whole spike is picked when one or to berries in the spike turn bright orange or purple. The berries are separated from the spikes and sun-dried for 7-10 days. When dried, berries retain characteristic-oriented appearance of black pepper of commercial use.

Besides India, pepper is cultivated in Indonesia, Malaysia, Brazil, Sri Lanka, Singapore, Thailand and few tropical countries. India enjoyed dominance till the fifties owing to production slack in countries like Indonesia. As on date, India shares only about 20.0 per cent of world pepper area, Malaysia, Brazil and Indonesia reporting a higher share. The average productivity in these countries is over 1000 kg/ha whereas in India it is only 263 kg. The area under pepper increased from 0.9 lakh ha in 1955-56 to 1.1 lakh ha in 1980-81 but the production has more or less remained constant at 0.29 lakh t. However, there has been a doubling in of quantity exported in this period. In 1981-82, about 70.0 per cent of total pepper production was exported.

Kerala accounts for about 96.0 per cent of total production followed by Karnataka (3.5 per cent) and Tamil Nadu. The productivity in Kerala increased from 225 kg in 1963-64 to 248 kg/ha in 1980-81, whereas in Karnataka it fell from 667 kg to 259 kg/ha. It may be pointed out that the yield level during the fifties was higher by over 12.0 per cent than in sixties.

In India, pepper is grown in homestead gardens as a mixed crop unlike the major pepper countries where it is grown as a mono-crop with improved cultural practices. Also, holdings are very small, the average being 0.7 ha and unviable.

In Karnataka, area under pepper increased from 2400 ha in 1963-64 to 3400 ha in 1975-76 but it declined to 2975 ha in 1980-81. Kodagu, Uttar Kannada and Dakshina Kannada are the major pepper producing areas. It is grown mainly as an inter-crop. During 1960-61, Dakshina Kannada was the major pepper producer (with 955 ha and 1026 kg/ha yield) followed by Uttar Kannada (920 ha and 175 kg/ha), Kodagu (552 ha and 462 kg/ha) and Shimoga (101 ha and 61 kg/ha). Over years the situation has changed in such way that Kodagu has emerged as the leading district (1072 ha) followed by Uttar Kannada (832 ha), Dakshina Kannada (748 ha) and Shimoga (222 ha). Though during 1976-77, area under pepper in the state was 3353 ha (Kodagu 1765 ha, Uttara Kannada 816 ha and Dakshina Kannada 456 ha), it declined to 2975 ha during 1980-81. After 1980-81, the area increase has been margina l with slight decline in yield rate in Dakshina Kannada. In this district, area under pepper is concentrated in the taluks of Puttur and Sullia. In all the major producing districts there has been a decline in the yield rate from 549 kg in 1960-61 to 59 kg in 1980-81 for the state, from 1026 to 237 kg in Dakshina Kannada, and from 462 to 291 kg/ha in Kodagu. In Uttara Kannada and Shimoga, however, there has been an increase in productivity but overing round the state average of 259 kg/ha in 1980-81.

The declining yield trend is visible especially since 1973-74. The decline is quite marked in Dakshina Kannada and Kodagu districts.

Pepper crop is subject to many pests (Pollu beetle, mealy bugs, thrips and top shoot borer) and diseases (foot-rot, slow wilt, and pollu disease). Proper control of pests and diseases, introducing and escalating the use of hybrid varieties, increasing plant density upto 1000 vine/ha, application of chemical fertilisers, etc, should enable an average yield of more than 1 kg of dry pepper per vine and the resulting competitiveness with other countries.

Cocoa

Cocoa (*Theobroma cocao L*), introduced in India around 1793 was brought under commercial cultivation only in the corrent century. Government restriction on its growing upto 1970-71 resulted in its poor performance (especially of criolla variety). Import and scarcity of cocoa resulted in an all time increase of price to Rs. 14 kg of wet beans in 1978-79. By 1980, cocoa area in the country increased to 20000 ha. The major purchase was made by Hindustan Cocoa Products Ltd. (Cadbury India) whose management lock-out resulted in the breakdown of the marketing system in 1980. At that time, Kerala State Cooperative Marketing Federation and Central Arecanut and Cocoa Marketing and Processing Cooperative Society were established. After a boom in cocoa prices, the cooperatives began offering Rs. 5.30 per wet kg of cacoa beans. The sudden fall in the prices and crop damage caused by rodent pests affected the interest of especially the small growers and interest in the crop began to subside. The present area is estimated to about 29000 ha with a production of about 5000 mt. Kerala accounts for 86.0 per cent of the area, the rest being shared by Karnataka (10.0 per cent) and Tamil Nadu. On an average, the yield of dry cocoa beans is 172 kg/ha though normal cases of 400 kg/ha are being reported. About 35 to 40.0 per cent of the crop is damaged by rodents. Other pests damaging the crop are mealy bugs, stem borer, stem girdler, aphids disease etc, and the diseases, black pod disease, canker, pink disease etc. Cocoa flowers from the second year of planting and the pods take about 140-160 days to mature and ripen. Normally 2 crops are harvested in a year. The crop is grown mostly in the interspaces of coconut and arecanut gradens as a mixed crop only. The experiment evidences show that the single edge cocoa planting is economically better than the double edge cocoa planting.

The launching up of Asia's biggest modern chocolate factory in Dakshina Kannada by the Central Arecanut and Cocoa Marketing and Processing Cooperative with a processing capacity of 9400 mt. of cocoa beans a year has given an impetus to its cultivation. Area under cocoa in Daskshina Kannada increased to 1967 ha in 1984-85 (about 47.0 per cent of the state total). Productivity is almost double that of Kerala. The slag in the cultivation of the crop may be removed by judicious mixture of coconut or arecanut, pepper and cocoa, followed by proper cultural practices and plant protection. The major incentive of course is improvement in the marketing system with help to small growers and possibilities of export improvement.

Rubber

There are about 35000 different articles made by rubber for various industrial, technological and domestic uses. In India, the annual turnover is about 3 lakh mt of which 67.0 per cent is in the form of natural rubber, 10 to 12.0 per cent synthetic rubber and the rest imported (both natural and synthetic). The above also includes reclaimed rubber accounting for about 10.0 per cent of total turnover.

Natural rubber is made from terpene hydrocarbon polymer obtained from rubber tree *nevea braziliensis.* It is a renewable resource depending on rainfall, soil nutrients, sunshine, etc. Synthetic rubber which is costly and made from petroleum derived chemical intermediaries is non-renewable. Recent technological advances in rubber production has resulted in increase in its area. During 1950-51, total area under rubber was 0.7 lakh ha with a production of 0.2 lakh mt. In that year imports accounted for about 21.0 per cent of production but increased to 48.8 per cent in 1960-61. However, with increase in indigenous production (1.53 lakh mt with an area of 2.42 lakh ha) the percentage of import was reduced to just 5.3 per cent in 1980-81. In recent years due to increasing, the proportion of import has been increasing (especially of synthetic rubber).

Kerala is the major rubber producing state followed by Tamil Nadu and Karnataka. In Dakshina Kannada district, rubber cultivation is concentrated in Sullia, Puttur, Belthangady and Kandapur taluks. Extension of rubber in non-conventional areas, adaptation into reforestation system, modernising small plantations, improvement of cultivation technology and encouragement of natural rubber production are some of the measures which could be adopted for development of rubber industry.

8

Summary and Conclusion

In Karnataka plantation crops contribute substantially to value of agricultural output. During 1970-71 the total share in gross value of output from agriculture was 18.4 per cent which increased to 20.5 per cent in 1979-80. It may be pointed out that during this period the share of agriculture to net state domestic produce declined from 60.8 to 54.0 per cent. The share of coconut which was 3.3 per cent of gross value of agricultural output in 1970-71 slightly increased to 3.7 per cent in 1979-80. The share of crops like cardamom, cashewnut and pepper increased substantialy over the base year.

Not only in value share but also in contribution toward tax revenue realised that the plantation crops have shown substantial improvement. The total amount of tax revenue realised from plantation crops increased from Rs. 46.4 mil in 1973-74 to Rs 89.6 mil in 1979-80. The share of coconut, cardamom, cashewnut and pepper in tax revenue increased from 10.5 to 18.6 per cent during the above period. From the export viewpoint, cardamom, cashewnut and pepper have fared better, though there is still import of coconut oil owing to poor oilseed performance. Among the plantation crops cashewnut has registered the largest increase in wholesale prices in the early eighties when compared to 1970-71 (30100) followed by tea (232.8), black pepper (229.9), cardamom (213.4) and coconut (183.0). The increase was more than that realised for cereals, cotton, sugarcane and fertilisers. Only pertroleum products and some items of machinery had registered a larger price increase. From the early eighties, the prices have further increased due to continuous drought and other national and international factors.

Coconut

India's share in world coconut area and production has been declining over the years. In states like Kerala, coconut yield has declined substantially. Though in Karnataka coconut has been performing well, the coastal districts fared poor. In Dakshina Kannada, the compound growth rates of area, production and yield during the past 20 years have been 0.4, 1.0 and 1.4 per annum respectively. In arecanut the growth rates have been 4.5, 8.9 ane 4.2 respectively indicating a better performance.

In Dakshina Kannada, coconut shares about 6.0 per cent of gross cropped area and the proportion is higher in the taluks of Kundapur and Udupi. Mixed type of gardens shared a higher proportion in both local and hybrid varieties. About 78.0 per cent of the trees were yielding. The proportion was slightly higher in non-paddy farms. The average yield per tree was 21.7 nuts in the paddy farms (3867 nuts ha) and 31.6 nuts in the non-paddy farms (5629 nuts/ha). On an average, 20.6 family mandays and 35.7 hired mandays were utilised per ha of garden. The proportion of family labour was higher in the smaller size groups. Organic manure to the extent of 940 kg and chemical fertilisers to the extent of 45.3 kg per ha were applied. The application of fertiliser was higher in the larger size groups. Total cost of cultivation in the paddy farms was Rs. 2323 and total income Rs. 7739 per ha. The net benefit-cost ratio was 2.34, the ratio being higher in the lower size groups. The cost of cultivation in the non-paddy farms was Rs. 2581 and income Rs. 11258 with a benefit-cost ratio of 3.37. Here also the lower groups reported a better ratio. On an average, 57.5 per cent of coconut harvested in the paddy farms and 65.2 per cent in the non-paddy farms was sold, the rest being disposed of in the form of tender coconut, home consumption and copra.

Arecanut

Unlike coconut-arecanut has been performing well in the country except in Kerala where the incidence of yellow leaf disease has been prevalent. Arecanut productivity has been increasing in states like Karnataka. In Dakshina Kannada, the crop shared about 6.0 of the gross cropped area like coconut. Taluks like Sullia, Puttur and Bantwal has substantial arecanut area. The proportion of pure type of gardens was higher but coverage under new strains was very negligible. About 76.4 per cent of the trees were yielding, the proportion being slightly higher in the case of paddy farms. Yield per tree was higher in the non-paddy farms (1.59 kg as against 1.34 kg in the paddy farms). Labour absorption was very high when compared to coconut (about 7 times higher). About double the quantities of organic manures and five times chemical fertilisers as that of coconut were being applied. The consumption of fertiliser was very high in the higher size groups especially in respect of non-paddy farms. Total cost of cultivation was Rs. 9159 in the paddy farms and Rs. 6998 in the non-paddy farms. The total income was Rs. 33670 and 39803 respectively. The net benefit-cost ratio in the paddy farms was 2.7 (very low for first group and very high for last group), higher than that reported for coconut. In the non-paddy forms however, the ratio was slightly lower (the lower size groups reporting a higher ratio than the average). For all the farms combined ratio in arecanut was 2.87, about 6.0 per cent higher than that for coconut.

Paddy

Paddy is an important crop in the district covering over 50.0 per cent of the gross cropped area, though the share has been declining over years as at the national level. Paddy productivity is slightly lower when compared to the state average. Excessive water-logging during rainy season and consequent low proportion of HYVs and fertiliser consumption are some of the reasons for lower yield rate. This is true for coconut also, though in arecanut the district comes nearer to the state performance. Kharif area is substantial followed by rabi and summer in order of importance. Except Sullia, all other taluks have substantial area under paddy. The sample farms utilised a larger proportion of hired labour. Labour absorption per ha was 147 mandays while that of animal labour was 40.5 days. Manure and fertiliser application was highest when compared to arecanut and other crops—1773 kg manure and 332 kg chemical fertiliser per ha. The smaller groups consumed a lower dosage of fertilisers (247 kg as against 407 and 378 kg in the III and IV groups). Total cost of cultivation was Rs. 4905 and income Rs. 12646 with a net benefit-cost ratio of 1.7, much lower than that for commercial crops. The unit cost was very high in group III followed by first group. Hired labour cost was responsible for this. The first group had a very low benefit-cost ratio (0.72) followed by the III group (1.58). The second group (1.86) and especially the fourth (2.04) reported higher rate of returns. About 25.0 per cent of the farmers cultivated paddy in summer also, but the benefit-cost ratio worked out to just 1.0. Increased expenditure on all types of labour and no manures and fertilisers but at the same time over 10.0 per cent reduction in yield rates were responsible for this lower rate of return.

When all the crops were combined for both the type of farms, the net benefit-cost ratio worked out to 2.77, the ratio being lower for groups I and III (2.36 and 2.60) and higher for groups II and IV (3.0 and 3.38). Groups I and III fared poor in paddy and arscanut and groups III and IV in coconut. In groups IV, because of better performance in arecanut, the coverage ratio was higher than others.

Other Crops

In cardamom, the variation in productivity has been quite marked, the general trend being decline in yield rate. Same is the trend in the case of black pepper also. Cocoa which started well in the beginning faces problems from pest and price. Rubber has had a favourable growth and its area has been increasing in particular areas replacing to a marked extent coconut, cashewnut and similar tree crops.

A major problem observed in the case of all the crops is inadequate or improper plant protection. For example, it was customary to find the farmers applying chemical protection measures only when extended symptoms of pests and diseases were noticed. Fortunately, the dreaded wilt disease in coconut and yelow leaf disease in arecanut had negligible incidence in the sample area.

Another major problem was unstable price over years for crops like arecanut and cocoa. Institutional financial arrangements have been effected in many commodities. The particular Commodity Boards also try to upgrade the viability and exportability of the concerned commodity. Whatever may be the facilities awarded or constraints present, the agro-climate of the region has to do a lot in effecting productivity and also the particular incidence of pests and diseases. It has been observed that due to increasing labour scarcity, the proportion of area under labour intensive crops like paddy and vegetables has been declining over years giving place to less labour intensive crops. Heavy rainfall distribution (and the resulting low irrigation facilities during the off season) affects crop yield not only for the particular season but also for the succeeding periods (as in the case of coconut).

Suggestions

A major activity which could be undertaken for the region is evaluation of water-tolerant high yield paddy varieties during monsoon. It has been observed that during excessive water regime, the land is not cultivated for a while. Extension facilities have to be strengthened especially in the case of plant protection measures. Even in manure and fertiliser application, proper dissimination of information becomes crucial especially during the peak cultivation periods.

From the viewpoint of banks, extended assistance is essential for crop growth and export from the district. The banks may adopt small holdings, providing technical and financial assistance for improved cultivation practices. The existing marketing methods have to be organised systematically. The banks can be assured of proper repayment of loans provided production and marketing are coordinated. Commodity Boards have to be strengthened especially for arecanut, for balanced forward marketing.

The Banks and Agriculture and Horticulture departments may select unviable cases and attempt to effect a balanced marketing system in order to strengthen the level of their organisation. Since prices of crops like coconut and arecanut vary more due to instability in production and input, a viable policy has to be initiated to increase and stabilise the yield rates. This is particularly so in the case of coconut. Increase in the production of coconut oil becomes essential by inter-arranging the different products and processes in coconut production.

Since plantation crops have commonalities in the particular region, a concentrated effort is required to develop all these crops along with the main crop which is paddy by proper institutional, infrastructural and financial incentives especially to small growers inducing a less labour and water intensive system.

9

Energy and Economics in Tea Processing

Introduction

Tea has two varieties—the small-leaved China (Sinensis) and the large-leaved Assam (Assamica) and is primarily a third world crop in India, Sri Lanka, Indonesia, Kenya, Argentina, Japan, U.S.S.R and other 20 countries. India has the largest area and accounts for the largest production, with highest productivity. But the share in world production has fallen from 65.7 per cent in 1950 to 39 per cent in 1980 and that in world exports from 47 to 26.2 per cent.[1] High cost of production and the export deorientation policy of the mid-seventies have been responsible for the low share in exports. But because of high qualities of Darjeeling and Nilgiri and the declining importance to the crop in some of the countries like Japan, this should invigorate the system which earns bulk of foreign exchange and provides direct employment to over a million of workers and yields the bulk of agricultural income tax. The productivity increase in tea, for example, has been quite impressive when compared to that of the green revolution crops.

Tea Technology

In green, semi-fermented or black tea manufacture age-old method of dessication prior to processing, slow cell rupturing before cutting, slow speed aeration to get desired degree of oxidation and slow drying to retain the inherent qualities of leaf have hindered modernization. The major changes in cultivation have been propagation of improved clonal culture, hedge planting, soil management, mechanical weeding, shading and intensive use of fertilisers. Somehow, mechanical plucking has not been successful due to lack of uniformity in the frame, space and also due to wastage. Similarly, with the case of mechanical pruning and cultivation. Hence, tea cultivation is predominently labour-intensive and plucking alone accounts for about 20-25 per cent of labour (in this case women as against men in tea processing). Quick replacement of uneconomic bushes and vacancies, proper pruning cycle and adherence to point plucking and processing help to maintain high quality and yield.

In processing, there have been changes in rolling from the initial orthodox system to the C.T.C (Crush, Tear and Curl) shredders, Rotorvane and recently Lawrie processor. In withering, the conventional tot system was displaced by withering troughs and the tunnel system. Fermentation has been converted into an automation process in C.T.C manufacture; also in the case of transportation of the leaf.

The introduction of C.T.C method in the fifties was a revolution in that, processing became a continuous process as against the batch process of the Orthodox system. The C.T.C process ruptures the cells of green tea leaf for more thoroughly and hence the drinking potential is almost doubled. The liquid from C.T.C. tea is thicker and brews quicker and the grains are smaller in size. A large percentage consists of fine and powdery grade (dusts). Orthodox tea is larger in size with the longest grade being one inch long broken orange pekoe, orange pekoe, orange fannings, broken pekoe, pekoe, souchang dust are some of the grades and the first grade accounts for about 30 per cent and the last grade for about 10 per cent.

Because of double the cuppage and less wastage in C.T.C. tea (but quality being low when compared to orthodox tea), the area under that process increased and by 1970, it accounted for about a half. In South India, the share of C.T.C tea increased from 4.1 to 44 per cent from 1961 to 1978, while for India as a whole, the increase in share was from 34.2 to 67.8 per cent. Orthodox tea showed a decline from 57 to 30.9 per cent (in India) and from 57 to 55.9 per cent in South India. The other two initial processes are legg-cut (since abandoned) and green tea (about 1.3 per cent in India). Increase in tea prices has made quality orthodox tea more remunerative and now the two teas are in a sort of balance, though new factories are encouraged to have most modern and C.T.C. machines. High grown tea is usually processed by the orthodox system to maintain quality, though yield may be less. Because bulk of Indian tea is grown on plains (75 per cent) and tea bushes mature faster on plains and greater the replacement, the higher the yield (new plants), C T.C. method does pose a problem to quality and export earnings.

Economics of tea manufacture

An important constraint in higher exports is the increasing cost of production of Indian tea and lack of marketing of instant tea and other new methods which other countries like China employ to capture the U.S. and other markets. Countries like Sri Lanka export bulk of the tea and quote lower international prices that it becomes difficult for India with Rs. 11.55 a kg (Rs. 6.86 by Sri Lanka, Rs. 7.92 by Bangladesh, Rs. 9.05 by Indonesia and Rs. 8.38 by Kenya in 1979) and only 14 per cent share in packet tea, tea bag and instant (in total production) to maintain the past trend in the present world of changing consumption styles—tea cans, lemon tea in bottles, carbonated tea, tea wine, cinnamon tea and other ready mix teas.

The unit cost of cultivation in large estates is lower than in the medium and small estates due to better organisation standard of cultural practices and efficiency of inputs.[2] The cost per kg of made tea (about one-fifth of green leaves) in 1963-64 was Rs. 4, in 1971-72 Rs. 5.50 and 1980-81 over Rs. 14.00. The detailed cost figure for 1979-80 and 1980-81 are presented in Table 9.1. Labour cost accounts for about half the total cost of made tea. Field input share is 8 per cent, while energy input share is about 10 per cent in 1980-81, registering slight increase over 1979-80. Selling expense share is about 7 per cent and when compared to the situation in the early sixties, the share of selling costs have remained more or less constant.

With the out-turn ratio of about 20 per cent of made tea to

green leaf (the moisture content varying from 74 to 82 per cent) the cost per kg of green leaf for 1980-81 would be Rs. 2.8. The cost of orthodox tea in some cases is 5-10 per cent higher than C.T.C and as already mentioned, the shift toward C.T.C. is one of higher investment capacity and quality considerations rather than cost as such. Because the tea soil has to be maintained acidic, application of zinc has been a normal practice. The intensive use of fertilisers according to yield level induces different return-cost ratios in the estates.[3] For example, the NPK application for a yield level upto 1800 kgs/ha is 180 : 40 : 90, where for yield level about 300 kgs/ha, it is 250 : 40 : 122 : 5 applied four times a year.

TABLE 9.1 : **Cost of Cultivation of Tea**

Cost item	*Description*	*1979-80 Rs/kg made tea*	*1980-81 Rs/kg made tea*
1. Inputs : Field	Fertilisers	0.56	0.78
	Plant protection	0.21	0.27
	Miscellaneous	0.09	0.08
2. Inputs : Factory	Fuel and power	0.79	1.10
	Machinery maintenance	0.18	0.18
	Miscellaneous	0.03	0.02
3. Selling Expenses	Packing Charges	0.55	0.65
	Marketing and transport	0.35	0.35
4. Building maintenance		0.22	0.25
5. Labour cost		5.09	6.18
6. Executive staff		0.33	0.38
7. Depreciation, tax and interest		1.91	2.47
8. Head office expenses (big estates)		1.01	1.04
Total Cost		11.32	14.01

Source; UPASI Tea Research Centre, Cinchona, Tamil Nadu,

The yield of made tea increased from 943 kg to 1489 kg/ha from 1950 to 1980 in North India as against 654 to 1752 in the case of South India. That of Tamil Nadu's was greatest (perhaps in the world) from 642 to 2009 kgs/ha during that period. The development cost for raising new plants is relatively lower when compared to that of replanting (3 : 4) and hence to fill in vacancies and raise new improved clones, heavy incentives have to be provided for this industry. In the Nilgiris in 1979, the proportion of bush area of above 40 years old the total was 44.6 per cent. If not profit as a share of paid-up capital is compared, (22 per cent in 1960 and 13.5 per cent in 1971), the problem acquires even more importance.

Though the proportion of packet tea is small, its higher cost is usually due to packing and other handling charges (Table 9.2). For packet tea, the additional expenses include wages, taxes, though the grocer's margin is less when compared to loose tea.

Energy in Tea Manufacture

After the green leaves (two leaves and a but) are plucked, soon they are transferred to the factory for physical or chemical withering. The leaves are spread at a rate of 10-20 kg/ha on a withering trough and artificially withered by warm air upto 24 hours to achieve 50-55 per cent wither in orthodox and 70-75 per cent wither in CTC, the moisture content being 45-50 per cent and 70 per cent respectively. Longer withers (especially in winter) increase the colour bu quality, development of flavour and greenness of tea liquor are reduced. It requires about 212.7 Kcal to wither one kg of made tea in CTC and 354.4 Kcal in orthodox manufacturing.

The withered leaf is rolled three times in the orthodox process for oxidization to enable the breakdown of the insoluble proteins into soluble amino acids. In CTC method, the leaf is cut to one roll and then through rotorvane machine crushed, teared are curled and necessary disintegration of the cells effected for fermentation. Fermentation is through spreading on floor on fermentation beds and held under high humidity in the orthodox process and in CTC method, it is through fermenta-

TABLE 9.2 : **Comparative cost of loose and packet tea**

	Loose tea Rs/kg	*Packet tea Rs/kg*
Selling price to consumers	18.77	21.94
Purchase price	14.42	13.69
Difference	4.35	8.25
Agents commission	0.08	—
Handling	0.06	0.60
Transport	0.25	0.60
Insurance	0.02	0.04
Loss in transit	0,02	—
Storage	0.02	0.07
Sales promotion	0.04	0.10
Interest, etc.	0.09	0.31
Miscellany	0.15	0.17
Establishment	0.12	—
Labour	—	1.68
Taxes	0.82	2.33
Others	—	0.89
Wholesaler's margin	0.53	0.28
Grocers' margin	2.14	0.78

Source : Domestic Marketing of Tea, National Council of Applied Economic Research, New Delhi, December 1979, p. 17.

tion drums. Rolling and fermentation consume about 6 times of withering time (about 3 hours) and after this, drying (firing) in chambers for about 20 minutes takes place. The black tea gets its familiar colour and then the leaf is sieved, graded and packed on chests of aluminium foils or plywood.

Orthodox rolling requires 72 horsepower for rolling and drying for a out-turn of 200 kg of made tea in hour. In the CTC process, the total power required is about 185 ph per 200 kg of made tea.[4] When all the processes are considered, the per kg and per ha consumption of energy turns out to be 3930.8 Kcal and 5.97 gigacal for orthodox and 4183.1 Kcal and 63.5 gigacal for CTC process respectively. (Table 9.3).

In the manufacture of green tea, there is no fermentation

TABLE 9.3 : **Energy use in tea manufacture**

(Kcal/kg of made tea)

Process	*CTC method*	*Orthodox method*
Withering	212.7	354.4
Rolling	592.0	190.1
Fermentation	27.0	27.0
Drying	3311.0	3311.0
Sorting and packing	40.4	48.3
Total	4183.1	3930.8
Per ha at 1519 kgs (10^6 Kcal*)	6.35	5.97

* For orthodox tea, if the yield level is lower, the corresponding energy use/ha will be larger. The yield level is average for all tea, 1980-81.

Source: Adopted from D.N. Barbora, 'Energy Management in Tea', *Tea Today* 1 : 17-21 (1981) and P. Raghavendran, 'Fuel Economy in Tea Manufacture', Bulletin No. 31. UPASI, Coonoor, 1974.

and after steaming, the leaf is alternately rolled and dried until it becomes too stiff for further manipulation. The energy required is about 20 per cent lower and in India, green tea accounts for about 1.5 per cent of total tea production, mainly for exports. Semi-fermented or coolong tea is manufactured with slight withering and fermentation and that of instant tea by spray-drying of black tea extract. The proportion of instant tea production in India is very insignificant. Apart from the factories attached to large estates, there are factories which process tea by 'bought-leaf' method wherein those who have no factories, process the leaf through bought-leaf factories.

Energy Use in South Indian Tea Industry

The material for this section is largely base on an UPASI study.[5] In 1981, South Indian Tea Industry used 830.7 teracal of fuel, of which firewood accounted for about 58 per cent followed by coal (20 per cent), oil (19 per cent) and leco (3 per cent). When compared to 1978, the notable change is decrease in the coal consumption and use of leco in 1981. Fuel costs per kg of made tea doubled in three years—from Rs. 0.14 to

Rs. 0.22 (firewood, from Re. 0.27 to Re. 0.58) Coal) from Re. 0.49 to Re. 1.03 (Oil). Firewood is the cheapest fuel. The use of firewood is smallest in Tamil Nadu and highest in Kerala and Karnataka.

Because of increase in coal prices, the proportion of local firing has declined from 34.16 per cent in 1978 to 19.94 per cent in 1981. Leco, though costly, because of its high thermal efficiency is slowly substituting coal in many factories. It is estimated that at a dependency ratio of 58 per cent in firewood, 20 per cent in coal, 7 per cent in furnace oil and 15 per cent in leco, total fuel energy required in 2000 A.D. would be 1120.0 teracal and per kg of made tea 5185 Kcal.

Electricity use in tea factories is mainly to run the machines and lighting and on average about 860 Kcal is required per kg of made tea.

Because of energy crisis, solar energy for drying, biogas for lighting, cooking and drying are being tried out as alternate energies. It is estimated that solar energy if fully developed and utilised would save about 38.8 per cent of energy used for drying in the Anamallais.[6] Under conventional drying, about 4996 Kcal of firewood energy or 5000 Kcal of leco and coal energy or 5005 Kcal of oil energy ratios could go down by 65 per cent in all the sources on hourly basis. Better energy management should save a considerable amount of wastage especially in drying, which consumes the bulk of energy used in manufacture.

Conclusions

If the raw is on environmental degradation and decreased energy efficiency, plantation crops act in two ways. For food crops the rate of utilisation of available energy is only about 60 per cent (especially in traction power and biomass). Misuse of energy resources would mean not only deceleration productivity but also of the environmental base. The technique of production of plantation crops is different in some respects. Ecologically, tea for example, is less energy demanding in the pre-processing stage. The nature of rainfall and long age of the plants necessitate dominance of chemicalisation and manual labour in pruning and plucking. It is both capital and labour intensive in the processing stage and hence per hectare energy use would be approximately 50-100 per cent higher (inclusive of processing). Whereas there is possibility for full use of say drought animal power by food crops, tea has to depend mostly on human labour and on purchase inputs. But cattle raising may be more ecologically suited to plantations and for the rise of small dairy plants and manure yields. It may appear as though a transfer of certain under-utilised natural resources to plantations would prove more beneficial to the environment as a whole. Also, better utilisation of forest biomass (with regenerating forestry) may solve fuel problems within the restrictive system of forest legislation.

If it is rationality that sways the decision analysis of small farmers it is to be better management of inputs and organisation which would enable the small estate owners to be economic planters. This would enable the industry prosper better and contribute a sizable share to agricultural growth.

Next, since processing is a must for final consumption, there may be wastages in that process. But for tea, only about 1-1.5 per cent of the green leaves are wasted when processed into black tea. For food crops, wastage in the field, post-harvesting and final stages of consumption is tremendous. In this respect, it is not a liability in development as some of the crops which refuse to grow but are so essential as to usurp the growth of (less) essential crops. Eucalyptus for example, might go well in plantation areas, though as a shade tree it is not suited. But when it comes to social forestry, it has to meet with numerous criticisms.

There is a tendency for high risk crops to be replaced by low risk crops, whether there is enough economic returns or not. In the Nilgiris, the area under potatoes which was 11,100 ha in 1969-70 has come down to 8370 ha in 1981-82 precisely due to unpredictability of the crop in periods of rainfall changes. And a part of its area has been taken away by tea. Some farmers my convert a bulk of potato area if tea is that riskless. But might also be a risk crop. Because forest which is the most unrisky is being done away with to meet raising population demand !

Tea cultivation would be labour intensive and as the vacant lands are filled with bushes and new processing machines installed, the productivity is bound to go up and depending on the international performance, Indian tea could maintain its lead even in the future. This should give a ray of hope in the presence of mixed performance of new agricultural technology.

Since tea manufacture is to strengthen our export market, product diversification to capture different life styles is the basic problem.

Notes

1. Due to this, India has emerged as the largest consumer of the king of beverages though per capita consumption is quite low when compared to great drinkers like U.K (which was leading some years ago) and Ireland. P.C. consumption increased from, .36 to .46 kgs and in U.K. it declined from 4.16 to 3.68 kgs during 1967 and 1976.
2. The productivity in small estates has been quite low (520 kgs/ha in below 5 ha as against 2015 kgs/ha in the above 400 ha farms in the Nilgiris in 1978. The recent development of cooperative tea factories try to organise the small growers (actually peasants) toward better productivity. It is reported that the recently started Government Tea Factory with the latest machinery and all that has given in initiation toward a new growth pattern.
3. Total number of estates increased from 6214 in 1951 to 13299 in 1979 in India; in the latter year, the Nilgiris accounted for about 6596 of them indicating the small size.
4. S. Ramaswamy 'Prospects of using Solar Energy in Tea Processing,' UPASI, Cinchona, 1983.
5. 'Energy for South Indian Tea Industry', *The Planters' Chronicle*, May 1983.
6. S. Ramaswamy, *Ibid.*

PART II

10

Some Problems of Rural Development

Introduction

The process of world economic development has witnessed the phenomenal growth of technology in various spheres and resulting in polarisation of habitats. Technical change became the major contributor to growth especially from the latter half of the last century in the United States, the United Kingdom and other advanced countries. In India, technology generation, transfer or adaptation remained ubiquitously absent in most sectors for long. After independence, technical cooperation with international agencies and governments enabled to build up the technology base especially in industries and slowly in other sectors. (Irrigation projects in the First Plan and heavy industries starting from the Second Plan).

Inadequate research and extension, capital formation and most of all, technology have compelled a widening disparity between the rural and urban areas in production and distribution. The urban areas, recently characterised by technological softwares are far ahead of the rural areas. The latter are characterised by low productive agriculture followed with even less productive village industries and transportation. Rural sector being the major partner naturally would demand a better share in the fruits of development, but owing to biased decision in favour of urban areas this has resulted in improve-rishing the rural economy more than that of the urban areas. It may be pointed out that the rural areas would lag far behind the urban areas in many development indicators for years to come. The poverty ratio there would be higher with the resulting under-employment, low productivity and inequality, besides their negative impact on all types of environment.

Approach of Rural Development

Rural development is a dynamic and long-term process to develop agriculture and other rural sectors and as such would demand a multi-pronged approach to achieve the desired goals of development. Institution building, appropriate technology generation, reduction in inequalities, poverty alleviation and sectoral development all these are inter-related and hence demand an integrative and coordinated approach for an all-round development of a self-subsistent village system.

The concept of rural development has undergone many changes starting from purely an economic criterion of increasing the low income level to a holistic one of provision of minimum needs, development low-priority sectors and population and so on. Specifically, development of the rural poor to bring them into the mainstream has been empasized in recent times. Starting from the notion of grow more food to feed the teeming millions and provision of community benefits to that of effective local participation for endogenous development, the app oach has taken many routes of varied objectives, tragets, areas, minimum packages, specialised programmes and development of weaker sections. The culmination into IRDP in the late seventies enveloping many individual programmes necessitated the demand for an integration of technology, environment and different sectors of the rural system, institutional and infrastructural development and provision of basic needs of food, housing, education, health and the like toward achieving the goals.

The IRD Programme has been criticised for its thin impact on the rural poor, undue emphasis on the primary sector, lack of decentralisation and coordination, bureaucratic delay and apathy and low investment profiles. The Project Evaluation Organisation of the Planning Commission and the National Bank for Agriculture and Rural Development, in their evaluation of IRDP, point out the weakness in identification of the beneficiaries, lack of infrastructural development and asset-push, lack of development of adjuncts, drawbacks in administrative and organisational structure, planning and implementation and lack of necessary monitoring and follow up action as contributing to the limited impact of the project. Because of these, the real poverty line remained substantially high especially when deflated to the rising price level. It was suggested that institutional agencies and especially banks should play a positive role in identification, target achievement (both quantity and quality) and financial assistance to the poor to raise their living standards. All these things would naturally demand a proper linkage between different sectors.

Persistence of poverty, inequality, burden of subsidies (these may be disincentives in certain respects), problem in planning and administration, etc. call for programmes to increase production and productivity, to develop resource income, credit and market facilities and to provide employment opportunities and minimum needs within a reasonable period of time. But the problem is so big and the remedial measures so small that its intensity remains more or less endemic.

However, there have been cases of successful performance of micro-level rural development projects in many areas over the years. But the success of the specificities rarely repeat at the macro-level. Hence the approach has to be one of aggregation of the micro experiences and local-level specifications. In this, a synthesis of working of so many agencies has to be striken. Intra-agency suspicion and supremacy will spoil the purpose of delivery system.

The Problems

The major problems in the successful functioning of the rural development programme have been meagre funds, unusually wide coverage, no uniformity in approach, high cost of administration, natural calamities, inappropriate infrastructural availability and use, lack of appropriate technology mix to cover all sections of the system, weak coordination and policy lapses and inadequate local participation. In short, the problems look so big as to thwart any small dosage of remedies.

The theory of 'trickle down' which was aimed at the initial process has not materialised. The green revolution technology did indeed increase farm productivity but poverty alleviation was not achieved. Many specialised programmes did not touch the base of the problem at all. Many projects have failed because of lack of understanding of local circumstances, institutional structure and sociocultural patterns. Wherever acceptable the beneficiaries have shown wide enthusiasm and adaptive capacity. Hence the projects have to be integrated with local resource endowment, institutional set-up, social services and facilities and the target group aspirations. The entails involvement of locals, training and effective extension and institutional assistance and so on.

Experience in other countries show the partial success of the rural development approach ; examples may be cited of Pakistan, Indonesia, Thailand, Philippines, Nepal, Burma and Sri Lanka. The East and South-east Asian countries fared better than the South Asian countries. The reasons are to be found in the deficient strategies and policies inconsistant with the objectives. Labour replacement, government intervention in market (biased toward urban technology) had negative impact in the poor countries. The role of subsidies vs incentives and infrastructure vs technology has to be considered here.

Sectoral Development

Of course, for an effective rural development, only a radical strategy with land reforms, endogenous technology in all sectors, involvement of poor, containing the socio-political consequence of increasing number of rural poor, and effective implementation are called for. Under the existing conditions, the process may involve greater role for decentralised decision analysis by coordinated group of agencies taking into consideration the local specificities, etc. Area development, specific sector development, relief, training and education, employment generation and other issues demand a fresh look at the appropriate involvement of the micro-level integration at the macro-level. The question is how the local system is affected by the influence of the external one (in this case government intervention and agency impact) toward self-sustainment.

Investment (both in capital and human resources) has been a major source of development (through technical change) and hence importance has to be given on the role of financial institutions in rural development. Proper identification of beneficiaries and areas of investment, provision of adequate financial and other incentives, effective coordination and follow-up, etc, will greatly change the present predicament. Nationalisation of banks, creation of regional rural banks, etc. have not made the commercial banks the leading partners in development. Multiplicity of agencies and the resulting under-coordination and disintegration have to be tackled through proper policy measures. Differential rate of interest, credit facilities to specific groups, extension, supply of inputs, price and market policies, administrative support and infrastructure development, etc. would not only make the role of institutions (especially banks) more effective but also appropriate in the present context of lop-sided development.

It may seem that a more plausible way toward rural development to be develop such of those sectors which have great potentialities and response to development. In regions where certain small village industries are concentrated it is better to develop such that the twin objectives of local resource utilisation and employment increase of local personnal are achieved. Bidi-rolling, cashew cultivation, sericulture, dairy, etc, are some of the examples of intensifying specific programmes (sectoral approach to rural development) which if developed would bring in direct and indirect rural development augmentation.

Mention may be made about crop development programmes coveraging many cereals, pulses, oilseeds, plantation and other commercial crops besides energy plantation. Here the area and target approaches may be combined with the sectoral approach for speedy rural development process. The recent cashew development programme is an instance where different approaches to rural development have been synthesised to develop not only the crop, but also the area, target groups and rural industries.

11

Cashewnut : An Introduction

Cashew (*Anacardium occidentale L*), native of South America, was introduced in West Africa and India by Portuguese travellers about 400 years ago. One theory was that the cashewnut tree was introduced to check soil erosion on hilly slopes and coastal lands. Thus it was raised as a soil binder and afforestation crop. The other reason floated by some is that the crop was introduced so that it could fill the cashew liquor demand of the western traders.

From Goa where it was first introduced, cultivation of cashewnut spread rapidly along the west and south-east coasts of the Indian sub-continent as well as the islands of Sri Lanka, Anadaman, Nicobar and Indonesia. In Asia, the conditions for cashewnut cultivation have been very suitable and it will grow up to an elevation up to 700 m. Now, cashew is grown almost throughout the tropics between the tropics cancer and capricorn.

Cashew is a low spreading evergreen tree reaching a height of 10-15 m. At around 8-10 years of age the plant reaches a normal height of 5-6 m and sheds its foliage and spreads over a radius of 4 m. If a plant is not properly trained in the early stages, it puts out the branches at the base and takes a bushy appearance. When there is a no-problem of pests, the stem is erect and the canopy spreading symmetrical like an umbrella shape. The lower branches will rest on the ground a few metres from the trunk and branches of older trees which have grown without being disturbed may creep over the ground over a considerable distance. When trees are harvested regularly, the lower branches are usually cut away to facilitate easy harvesting. The economic life of a cashew tree is about 30-40 years. The yield will to up from 50 kg per ha in the 4th year to

100 kg per ha in 5th year, gradually increasing to over 1,000 kg per ha after the 10th year.

Cashew is grown on a wide variety of soils. The soils may be deep friable, well-drained, sandy loan soils without a hard pan with the phrootic water level at a depth of 5-10 m. Cashew cannot with stand a bad drainage, stagnating water and flooding. It can be grown under very low rainfall ranging from 300 mm to high rainfall of over 4,000 mm. It has been proved that *A Occidentale* is the only species which has been introduced outside the new world. Within Central and South America as many as 20 species of cashew are known to exist. Systematic cultivation has received attention only very recently and the available statistics on area and production are conflicting and are of doubtful reliability. Even today most of the cashew collected and processed for World Market are picked from selfsown wild growing trees. The World Bank has estimated that only about 2-3 per cent of the annual world production is from systematically planted plantations and bulk of the crop originates from wild growth in small holdings.

Crop Management

Cashew tree produces innumerable flowers of which less than 10 per cent are perfect. Under normal conditions, nearly 85 per cent of the flowers are fertilized. But only 4-6 per cent of this reaches maturity. The maximum shedding of the unproductive flowers takes place during the mustard, peanut and later stages. Heavy wind, frost, insect attack, etc., are some of the factors for low productivity. The fruit of the cashew tree is grey coloured, kidney shaped and borne on the fleshy thick hypocarp. It ripens fully in 60 days. The false fruit (also called the cashew apple) is juicy and sweet when ripe. It varies in size, colour, juice content and taste. Usually the fully riped apple is 10 times heavier than the nut. It is a rich source of vitamin C and sugar.

The nuts vary in size. shape and shelling percentage. The single nut is 3-5 cm in length and 2 to 3.5 cm in width. The weight of a single nut usually varies from 3-20 gm and the shelling percentage varies from 15-30.

The edible and nutritious kernel contains fat, protein,

carbohydrate, minerals and vitamins. Cashewnut shell liquid is an important raw material in paint, chemical and wood industry. Cashew apple can be utilised for production of beverages, juice, syrup, and jam. etc., Cashew tree is a valuable source of raw material for pulp industry and is a major source of fuel in many places.

The nut contains about 21.0 per cent proteins by weight, 47.0 per cent fat and 22 per cent carbohydrates. This is rich in iron, calcium and phosphours. It contains many active amino acids.

Cashew apple contains upto 5 times vitamin C as that of citrous fruit. The apple is eaten as such by sucking juice and discarding residual fibrous matter. Sugar or salt is added sometimes to reduce astringency. Steaming the cashew apple is most efficient method in removing the astringent and acrid qualities. From cashew apple, clarified juice, cloudy juice, syrup, juice concentrate and carbonated beverages can be prepared. Vinegor, candy and jam, chutney and pickles can also be processed.

Cashewnut shell liquid is valuable raw material for a number of polymer based industries like paints and varnishes, resins, industrial and decorate laminates, brake linings and rubber compounding resins. The CNSL is extracted from the raw nuts by hot oil bath, expeller, kiln and solvent processes.

Cashew in India

India has the largest area and is the largest producer of raw cashewnut and exporter of cashew kernels in the world. Right from the start, the industry is mainly export-oriented and highly labour intensive, employing over 2.5 lakh employees. In the world, Tanzania, Kenya, Mozambique and Brazil account for the bulk of total cashew production after India. The total cashew production during 1972-73 was 3.75 lakh t, of this, the share of India was about 20.0 per cent, one of the leading one. During 1978-79, it is seen that over 1/3 of total production is accounted by India. During 1981-82 also the share is similar. During 1982-83, world cashewnut production, however, declined to 2.78 lakhs t, and India's share was about 43.0 per cent. India is maintaining the lead in cashew production through substantial imports of raw cashewnut from the African countries and exporting the processed nuts. Tables 11.1 and 11.2 describe the world cashewnut production, export, import, indigenous production respectively in India. It is seen that the import of raw-nut has been considerable during 1960-61 to 1975-96. For example, the import was 1.37 lakh t during 1975-76, but in the subsequent year it declined to 0.75 lakh t and by 1980-81 to a mere 0.26 lakh t. Export has been considerable in the early seventies (0.66 lakh t in 1972-73 and 0.65 lakh t in 1974-75) but by 1980-81 it could be seen that export has been declining and in that year it stood at 0.33 lakh t. Competition from other countries and resulting decline in raw cashew nuts, improved processing facilities in countries like Brazil and Mozambique, and declining productivity trends in India all these necessitate a new emphasis on cashew development.

Cashewnut is one of the important foreign exchange earners fetching over Rs. 150 cr per annum. During 1970-71, the total area under cashewnut was 2.893 lakh ha and increases to 5.02 lakh ha in 1984-85. For an increase of over 70 per cent in area, the increase in production has been marginal. During 1970-71, total production was 1.77 lakh t, giving a yield rate of 604 kg per ha. In the subsequent years, the yield level has declined to 606 kg per ha in 1973-74, 428 kg in 1976-77, 407 kg per ha in 1981-82. During 1984-85 the estimated yield rate has been about 520 kg per ha, indicating increasing trend from now onwards. Table 11.3 outlines the statewise area, production and yield trends from 1970-71 to 1984-85. Among the states, Kerala accounts for the bulk of area and production. During 1970-71, it accounted for about 37.0 per cent of total area and 65.0 per cent of total production. This shows that the yield rate is also substantial in that state. Tamil Nadu, Goa, Andhra Pradesh, Maharashtra and Karnataka and substantial area under cashewnut, in order of importance. In yield rate, however, the position was Karnataka, Andhra Pradesh, Tamil Nadu, Maharashtra and Goa in order of importance. During 1975-76, the yield rate in other states except Kerala has declined. In Karnataka for example, it has declined from 933 kg per ha in 1970-71 to 427 kg per ha in 1975-76. During 1981-82, the yield rate in Kerala has declined to 801 kg per ha and in Karnataka, the yield rate has been

308 kg per ha. In area, the share of Kerala has been 30.6 per cent. Karnataka has improved its share in area along with Orissa in the terminal year. Kerala however is maintaining the lead in productivity rates. During 1984-85, share of Kerala in area has been 30.2 per cent, that of Tamil Nadu 18.9 per cent and Karnataka and Andhra Pradesh each about 12.5 per cent. There is a slight improvement in yeild rate 816.0 in Kerala, 345 in Andhra Pradesh, 303 in Karnataka, 151 in Tamil Nadu and 546 kg per ha in Maharashtra.

An on today, India shares about 42.0 per cent of the total world cashew production. Though India exports cashew kernels, it does not produce enough raw nuts to meet the complete requirements of the processing industry, which comes to about 4.5 lakh t. Therefore, India has to import substantial quantities (up to 2.5 lakh t) mainly from Kenya and Tanzania and Malogacy Republic. Development of cashew processing industry in east African countries particularly, Tanzania and Mozambique and emergence of Chine and Brazil as cashew exporters have resulted in gradual drying up of the raw nuts available for India. Therefore, the bulk of the requirements of the processing industries has to be met by improved internal production. This highlights the necessity to improve the productivity of this crops and also its export performance.

Cashew in Dakshina Kannada

In Karnataka cashew is grown over an area of about 63,000 ha with a production of 190, 100 mt of raw nuts during 1984-85. During 1960-61, the area was 6,388 ha which increased to over 39,000 ha in 1978-79. Table 11.4 outlines the area under cashew in different districts from 1960-61 to 1978-79 and yield per ha during the terminal year. It is seen that Dakshina Kannada accounts for the bulk of the area (it was 77.0 per cent during 1960-61 and 1978-79). After this period other districts like Shimoga, Belgaum, and Uttara Kannada have increased their share of cashew area. During 1986-87, the total area in the Dakshina Kannada district has been 29,091.1 ha with a production capacity of over 14,000 mt of raw nuts. When we compare to the 1979-80 level when the productivity was about 700 kgs per ha, it is seen that the yield rate in 1986-

87 has been low at about 500 kg per ha. This is due to the fact that inspite of the increase in area during the World Bank Project period (1980-81 to 1986-87) the additional area brought under cashew cultivation has not started yielding returns. It is commonly after the third or fifth year after planting that the trees start yielding. This factor should be kept in mind when we evaluate the performance of the Project immediately after its completion.

In Dakshina Kannada; there is vast scope for expanding cashew cultivation in the waste lands and mountain slopes. An estimated area of 61,395.56 ha is available as waste land, of which some 10,000 ha has been brought under cashew cultivation during 1980-81 and 1986-87

Under this Project, a large scale enhencement of area has been implemented after 1980-81. About 10,000 ha spread over five year has been covered under this Project. The scale of loan given has been Rs. 2,805 per ha of which Rs. 900 forms the subsidy portion. Apart from this, financial assistance has been also extended to purchase sprayers. The loans have to be advanced to the farmers through primary land development banks and commercial banks at the nearest point. Over 8,300 small holdings with an area of about 10,000 ha have bcen covered during the Project period.

Under the centrally sponsored subsidy scheme for cashew cultivation, provision has been made for conducting fertiliser and plant protection demonstrations, particularly to control tea-mosquite. The Department of Horticulture is procuring seed nuts from high yielding cashew from the research stations which are supplied to the beneficiaries at their door. So, 2.5 million seedlings have been planted in the small holder plots. But this in itself is not an end to increase production. There are uneconomical and non-yielding trees with different age groups and the most important task is to improve the present yield level by adopting latest techniques.

Cashew plantations are raised in marginal lands where no other crop could give and economic return and it receives very little attention from the growers. The reason for the low production consisting of selfsown seedlings and poor management. Scientific management has become imperative to increase cashew production. Manuring, cultural operations, weeding, mulching, cover cropping, pruping and plant protection measures are some of the aspects to be consider for improving the productivity.

Regular fertilizer application to ensure early and higher yields to new plantations and regular high yields from immature trees has became necessary. It has been resorted that a thirty year old cashew tree removes 2.80 kg N, 0.75 kg P_2O_5 and 0.75 kg K_2O per year. Preliminary trials on nutrient requirements indicated that an annual application of 500 g N, 125 g P_2O_5 and 125 g K_2O per tree per year would be the optimum dose for cashew, to be applied in two equal split doses first in pre-monsoon and the rest in post-monsoon when there is optimum moisture in the soil.

Coverage of the World Bank Project

Since cashew plantation is in the small and unorganised sector, importance has not been given to improvement in cultivation over years. Local research, the Horticultural department and the Karnataka Cashew Development Corporation have been trying to invigorate the production of cashewnut, especially at a time when the cashew yield has been declining and international competition has slackened India's chances of leading the export market. The major problem was that the cashew processing industries, hitherto dependent on import of rawnuts from African Countries, were forced to work below capacity utilisation on the consequent of decline in imports. The Karnataka Cashew Development Corporation chalked out of programme to bring additional area under new plantations and rejuvenate the existing ones, with a purpose to increase cashew production to keep the processing factories in work. Also, ushering in of new plantations would mean adopton of improved varieties through which the yield rate could also be increased.

The objective of the cashew development programme in private holdings was to increase the total cashewnut productions by raising the productivity in a certain proportion of the existing plantations and to bring in about 10,000 ha of new area under cashew plantations. This is to enable increase cashew production by the thirteenth year of the project. Since Dakshina Kannada is a district blessed with favourable agro-climate for growth of cashew and since the processing factories are facing shortage for raw cashewnut, increase in cashewnut production becomes inperative. Hence the World Bank Project was launched since 1980-81.

Objectives and Methodology

With a physical target of covering about 10,000 ha under cashew cultivation and with a financial outlay of Rs. 3.47 crores (raised to Rs 4.46 cr), the World Bank Project could be evaluted as to its performance and efficacy. The following are the major objectives of the study :

(i) to study the working of the project and how the various targets have been achieved;

(ii) to study the cashew cultivation practices in the beneficiary cases and to find out the extent to which the new methods of cultivation are being adopted;

(iii) to find out the difference of adoption of practices between the beneficiaries and the non-beneficiaries; and

(iv) to study the relative importance of cashew in the farm economy and to study production, and marketing procedures and the related issues.

The sample area selected has been the Dakshina Kannada district, where the World Bank assisted India cashew project in being implemented. All the eight taluks of the district have good coverage under cashew and six taluks have been selected for collecting the primary data on cultivation and other practices. The six taluks are Mangalore, Bantwal, Puttur, Belthangady, Udupi and Karkal.

About 150 beneficiaries who have been receiving the subsidy under the Project have been intervived in the above 6 taluks. About 100 non-beneficiaries were also interviewed for comparison with the subsidy cases. A questionnaire was convassed amongst the farmers in different regions and the field investigators contacted the farmers and collected the necessary information. The questionnaire contained information on cultivation practices, subsidy received, adoption of improved methods, performance of cashew vis-a-vis other crops, marketing and other particulars. Cashewnut is harvested during March-April, so the cashew season starting from May 1985 to April 1986 was covered for cendysis. The field study was undertaking during April to August 1986 in various places.

A major limitation of the study is that since the fruits of the Project would be realised only when the first batch of the new plantations would start yielding the nuts. This has just started, but is not the optimal stage anyway. Though the efforts of Horticulture department, and cashew research stations to propagate new cashew varieties and improved methods have resulted in some improvement, from the farmers' viewpoint, a stage has not yet resulted wherein one could evaluate the new methods per se. Since cashew is still not given an important place in the household's technology, this evaluation study has to keep those limitations towards analysing the data collected from secondary and primary sources.

TABLE 11.1 : **World raw cashewnut production**

('000 mt)

Country	*1961-65*	*1971*	*1981*
World	368.0	582.7	454.3
India	145.2 (39.5)	217.0 (37.2)	190.0 (41.8)
Mozambique	131.2 (35.6)	200.0 (34.3)	95.0 (16.5)
Tanzania	64.4 (17.2)	120.0 (20.6)	72.3 (15.9)
Brazil	11.9	20.0	85.0
Philippines	6.0	5.4	4.0

Note : (Figures in Parentheses indicate percentage share)
Source ; *FAO Yearbook 1982*, FAO, Rome, 1983,

TABLE 11.2 : **Indigenous production, import and export of cashew kernels from India**

('000 mt)

Year	*Export of kernels*	*Rawnut equivalent @ 24% recovery*	*Import*	*Indigenous production*	*% to export*
1946-47	15.4	64.2	31.6	32.7	50.9
1950-51	26.3	101.3	54.8	46.5	45.9
1955-56	31.4	130.7	63.1	57.5	51.7
1960-61	43.6	181.8	118.3	63.5	34.9
1965-66	51.8	213.6	160.6	53.0	24.8
1970-71	50.3	209.5	169.4	40.2	19.2
1975-76	53.6	223.5	137.2	86.3	38.6
1980-81	32.8	136.7	25.7	110.7	81.2

Source: L. Krishnaswamy, 'Production and Supply of Cashew', *Cashew Causerie*, Vol. III, No. 4, 1981.

TABLE 11.3 : **Statewise Cashew area and Production**

(Area in '000 ha and yield in kg/ha)

State	1970-71		1975-76		1981-82		1984-85	
	Area	*Yield*	*Area*	*Yield*	*Area*	*Yield*	*Area*	*Yield*
1	*2*	*3*	*4*	*5*	*6*	*7*	*8*	*9*
Kerala	107.5	1073	109.2	1092	147.4	801	151.4	817
Karnataka	17.6	939	35.7	427	58.3	309	63.0	308
Andhra Pradesh	25.8	504	31.0	406	57.3	349	64.4	348
Tamil Nadu	81.9	262	96.8	108	94.8	111	94.8	151
Goa	34.5	162	39.8	142	44.1	193	9.6	—
Maharashtra	17.7	198	33.4	126	22.7	441	22.7	546
Orissa	5.3	236	21.9	178	48.6	165	53.5	187
India	292.7	604	308.7	555	481.0	407	502.0	420

Source: Directorate of Cashewnut Development, Cochin.

TABLE 11.4 : **Area and yield of Cashew in Karnataka**
(Area in ha and yield in kg/ha)

District	*1960-61*	*1965-66*	*1970-71*	*1978-79*	
				Area	*Yield*
Bangalore	94	123	227	199	90
Kolar	208	256	587	773	98
Shimoga	24	141	2,651	6,018	53
Tumkur	25	12	34	220	59
Chitradurga	21	21	—	9	—
Belgaum	9	5	19	1,464	97
Uttara Kannada	719	675	962	1,556	17
Chikamagalur	26	133	65	159	75
Dakshina Kannada	4,905	7,665	12,547	17,436	83
Hassan	—	1	5	5	—
Kodagu	329	611	497	932	146
Mandya	—	2	—	126	47
Mysore	2	286	207	37	27
State	6,388	9,946	17,821	38,534	77

Source : Directorate of Economics and Statistics, Bangalore.

12

The Project

Introduction

Under the World Bank Project, due consideration has to be given for the development of cashew crop per se and in relation to other crops. The project in one of the schemes under which increase in new cashew plantations and improvement of existing ones has to be accomplished. The scheme was sanctioned with the assistance of International Development Agency through the National Bank for Agriculture and Rural Development. The area of the project was to be 2565 ha and the project period from 1985-86, the plantations raised during 1980-81 to 1980-86 have to be maintained for at least 5-10 years for realising the yield. Total cost for 10 years was estimated to be Rs. 8.2 crores which was split as below : maintenance cost Rs. 4.2 cr, staff salaries Rs. 2.4 cr and operating expenses Rs. 1.6 cr. Sources for this would be Rs. 7.2 cr through sales revenues, Rs. 1.4 cr through loans and Rs. 0.7 cr through equities. The ratio of equity and loan component has to be maintained as 1.2. Hence the equity of the state government would be Rs. 0.7 cr and the loan through NABARD Rs. 1.4 cr. The anticipated revenue from cashewnut would be Rs. 7.2 cr. It was estimated that the loan instalments would be for a period of 15 years. The major centre of field management would be the plantation managers situated nearer Puttur and Kundapur. The Karnataka Cashew Development Corporation, an independent Corporation has also taken loans through NABARD for improvement of older plantations.

The production programme involved 1000 ha of replantations. The area selected for raising plantations are the understocked old plantations belonging to forest department and Karnataka Cashew Development Corporation. The seed for raising the nursery would be obtained from clonal or chard of KCDC, Cashew Research Station, Ullal and CPCRI, Vittal and Shanthigodu. The areas were to be suitably developed by removing the existing unvarirably growth. Seedling progenies raised in plastic bags would be adopted in pits of 50×50×50 cm. The plantation areas are to be fenced by digging cattle proof trenches and putting barbed wire over the fence. Improved agro-techniques such as fertiliser application and pesticide spray besides doing soil woking around the plants, weeding and mulching were to be followed. For conservation of moisture and soil, inward terracing with trap trenches were to be advocated. A subsidy of Rs 500 per ha would be given by the government of India and Karnataka for new plantations. The project would be implemented over a period of 5 years upto 1985-86.

The Targets

The Physical Target of the Project has been as follows :

Year	*Physical Target* (ha)
1980-1981	338.0
1981-1982	1424.0
1982-1983	1450.0
1983-1984	3392.0
1984-1985	3300.0
Total	9,904.0

Total cost for 5 years estimated at Rs. 227.0 lakhs for raising new plantations has to be distributed as follows :

1.	Planting improvement	Rs. 82.0 lakhs
2.	Staff Salaries	Rs. 36.0 ,,
3.	Civil Works	Rs. 37.0 ,,
4.	Vehicles and equipments	Rs. 15.0 ,,
5.	Operating expenses	Rs. 27.0 ,,
	Total	Rs. 227.0 ,,

Table 12.1 explains the area targeted for new plantations

and for improvement of old plantations for various taluks of D.K. according to different revenue firkas. It is seen that the area targeted for new plantations is 2500 ha and for improvement another 500 ha from 1980-81 to 1984-85. The incremental production cost for new plantation is presented in Table 12.2 wherein for the 2500 ha plantations (from 1980-81 to 1984-85) the expenditure for upto 1990-91 is estimated at Rs 26.145 lakhs. Table 12.3 presents the incremental production cost for the improvement of 500 ha of cashew plantations estimated at Rs. 5.135 lakhs.

The Project was extended by two more years initially and then beyond September 1987 with an additional target of 2000 ha, thus bringing the Project target area to about 12,000 ha of cashew plantations.

Extension

Under extension programme, propagation of new Ullal varieties of improved cashewnut as developed in the Argicultural Research Station (UAS) Ullal has been undertaken. Improved cultural practices were to be advocated for better cultivation methods. Important in these are adoption improved varieties, planting space and method, fertiliser application, plant protection measures, soil and moisture conservation practices, digging pits, etc. For this yearly subsidy besides the loan would be made available to the farmers.

Training of cultivators who are project beneficiaries at ARS, Ullal to import the technique to rejuvenate the unyieldy and uneconomical cashew trees is being given. The new technique on top working evolved by the ARS, Ullal has also to be imparted. The training programme envisages to depute 150 cultivators from each taluk during 1986-87. One day training in this regard were to be given to improve at least 12,000 trees. Besides training, the Project envisaged establishment of a scion bank/bud wood nursery with selected clonal materials for mass multiplication of the planting material. Also raising of soft wood grafts has also to be undertaken.

The calender of operations has been recommended as below :

Month		*Operation*
December-January	:	Spraying with Endosulfan weeding.
February-April	:	Harvesting and collecting of nuts, nursery work.
May	:	Fertiliser application weeding, bunding, etc.
June-July	:	Planting of Seedling, die-back affected trees to be sprayed.
August-September	:	2nd dose of fertiliser application.
October-November	:	Spraying.

The schedule of fertiliser application is shown in Table 12.4. The dosage is per tree per year in grammes of nitrogen, phosphorous and potassium. During the first year, a total quantity of 100 gms of nitrogen and 80 gms of phosphorous (in two doese one before monsoon and the other after the monsoon) has to be applied per tree. The quantity of nitrogen would double in the second year and quadruble in the succeeding year. From the fourth year onwards about 500 gms of nitrogen, 120 gms each of phosphorous and potassium have to be applied per tree per year in two equal doese. But intensive fertiliser application has not yet become fully incorporated into cashew cultivation as would be seen later. Even in some cases where it is adopted, it is in a haphazard way.

Mention may be made about the occurrence, damage and intensity of infestation of certain cashew pests on cashew cultivation. Table 12.5 describes the particulars of 21 pests, the nature of damage caused by them, their seasonal occurrence and the intensity of infestation. Of the pests, plocalderus ferrugineus (stem and root borer) macalla moncusalis walker and orthaga exrinacea hamps (leaf and shot webling caterpillers) acrocercops syngramma (leaf miner) and helopeltis antonii (tea mosquito bug) are some of the major pests which attack the cashew saplings and trees. The pests may be more vigorous in attack during particular reasons only. Thus the intensity of infestation of the stem and root borer is very serious during the season May-July, moderate during August-September and mild during October-April. In cases of sever infestation, the tree is killed. Some are regular pests, while other are minor or occasional. Unless the pests are checked, the cashew tree growth may be affected and hence the yield productivity would also greatly decline.

The tea mosquito bug may be controlled by spraying with endosulfan of quinalfos 0.05 per cent as high volume spray or 0.1 per cent low volume spray at the time of emergence of new flushes (October) and of the inflorescence (December-January). Further, plant protection measures may control the pest and also insects associated with fruit drop. Apart from spraying pesticides and insecticides adoption of phytosanitary measures like removal of dead trees are also important in control of pests. Removal of dead branches and pest-affected tree would also be a right controlling measure. Damping off of seedlings under water-logged conditions and other diseases control requires spraying of bordeau mixture, pruning of affected branches, protecting the cut surface, etc.

It may be pointed out that when the recommended dosage of plant protection chemical application is considered, most of the farmers are yet to adopt the same. There may be many reasons like non-availability of chemicals, sprayers, and skilled labour, price and other cost economic factors.

TABLE 12.1 : **Phasing of the Cashewnut Plantations Programme.**
(Area in ha)

	Year				
	1980-81	*1981-82*	*1982-83*	*1983-84*	*1984-85*
			New Plantation		
Kuntur	40	360	—	—	—
Nelodomale	—	25	175	—	—
Petroma	—	—	275	125	—
Alchimanmale	—	—	—	200	200
Hulimeshepore	—	—	300	300	—
Gunduberu	—	—	—	125	175
Janalane	200	—	—	—	—
Total	40	585	750	750	375
Improvement of Existing Cashew Plantations					
Vittal Mandur	—	35	75	—	—
Sarve Mandur	25	65	75	26	—
Ajjibettur	—	—	—	—	—
Eajore	—	—	—	115	—
Kalthod	—	—	—	—	84
Total	25	100	150	141	84

Source : Karnataka Cashew Development Corporation, Mangalore.

TABLE 12.2 : **Incremental Production Cost (New Plantations)**

(Expenditure Rs. in lakhs)

Year of plantation	*Area to be planted (ha)*	*1980-81*	*1981-82*	*1982-83*	*1983-84*	*1984-85*	*1985-86*	*1986-87*	*1990-91*
		1	*2*	*3*	*4*	*5*	*6*	*7*	*8*
1980-81	40	0.520	0.278	0.368	0.328	0.380	0.400	0.410	0.440
1981-82	585	—	7.605	4.065	5.382	4.680	5.557	5.850	6.435
1982-83	750	—	—	9.750	5.212	6.900	6.000	7.125	8.250
1983-84	750	—	—	—	9.750	5.212	6.900	6 000	7.875
1984-85	375	—	—	—	—	4.875	2.610	3.450	3.843
Total	2,500	0.520	7.883	14.183	20.672	22.047	21.467	22.835	26.845

Source : Karnataka Cashew Development Corporation, Mangalore.

TABLE 12.3 : **Incremental Production cost improvement to the existing plantations expenditure.**

(Rs. in lakhs)

Year	*Area to be improved in ha*	*1980-81*	*1981-82*	*1982-83*	*1983-84*	*1984-85*	*1985-86*	*1986-87*	*1989-90*	*1990-91*
		1	*2*	*3*	*4*	*5*	*6*	*7*	*8*	*9*
1980-81	25	0.375	0.262	0.262	0.250	0.252	0.250	0.250	0.262	0.262
1981-82	100	—	1.500	1.050	1.050	1 000	1.010	1.000	1.030	1.050
1982-83	150	—	—	2.250	1.575	1.575	1.500	2.250	1.530	1.545
1983-84	141	—	—	—	2.115	1.480	1.480	1.410	1.410	1.438
1984-85	84	—	—	—	—	1.260	0.882	0.882	0.840	0.840
Total	500	0.375	1.762	3.562	4.990	5.567	5.122	5.792	5.072	5.135

TABLE 12.4 : **Schedule of fertiliser application**

(gm/tree/year)

Period	*May-June*			*September-October*		
	N	P_2O_5	K_2O	*N*	P_2O_5	K_2O
I Year	50	40	—	50	40	—
II Year	100	40	30	100	40	30
III Year	200	60	60	200	60	60
IV Year	250	60	60	250	60	60

Note : 1 g N = 2.1 g urea.
1 g P_2O_5 = 6 g superphosphate.
1 g K_2O = 1.6 g muriate of potash.

TABLE 12.5 : **Seasonal occurrence, damage and intensity of infestation of certain cashew pests**

Sl. No.	*Name of the pest*	*Damage*	*Seasonal Occurrence*	*Intensity of Infestation*	*Remarks*
1	*2*	*3*	*4*	*5*	*6*
1.	Plocaederus ferrugineus L.	Bores into stem and roots, feeds on internal tissues and kills the tree in severe infestation	May-July Aug-Sept. Oct-April	Very severe Moderate Mild	Overlapping generations throughout the year were observed.
2.	Indarbela tetranis M.	Attacks branches, bores holes into stem, feeds on internal tissues thus drying of twigs.	June-Sept. Oct-May	Moderate Mild	
3.	Macalla Moncusalis Walk.	Webs on tender shoots and leaves and later spreads to fruits and nuts and cause damage	June-Sept. Oct-Jan. Feb-May	Severe Mild Service	A regular occurring pest.
4.	Orthaga exvinacea M.	Feeds on leaves highlying in webs	Aug-Oct.	Mild	Occasionally serious pest.

1	2	3	4	5	6
5.	Holotricha sp.	Adult defoliates the top branches. Grubs feed on roots of plant, cause wilting and drying of entire plant	June-July Aug-Oct. N ov-May	Moderate Severe Mild	Nursery and young plantations suffer more.
6.	Thallosodes guadraris Guen.	Settles and feeds on margins of tender leaves	Oct-Dec. Jan-Sept.	Moderate No Incidence	Minor incidence
7.	Bombetelia reacosatrix Guen.	Feeds on tender leaves by making small holes all along	Oct-Dec. Jan-Sept.	Moderate No Incidence	Defoliater
8.	Metanastria hyrtaca Cram.	Defoliation	Jan-Aug. Sept-Dec.	No incidence Moderate	Sporadic
9.	Acrocercrops Syngramma M.	Mines the young leaves causing white blistered patches.	Jan-April May-Aug. Sept-Dec.	Law incidence Moderate Severe	Regular pest
10.	Rhipiphorothrips rothrips Cruentatus H.	Congregates and sucks the sap from the leaves	March-May June-Feb	Severe Mild	Occasionally Serious
11.	Rhynchothrips rapensis G.	Sucks the sap from inflorescence resulting in dried panicles.	Feb-April May-Jan.	Severe No incidence	—
12.	Myllocerus sp.	Attacks shoot tips, tender leaves and panicles.	Jan-June July-June Oct.-Dec.	Moderate Severe Mild	—

1	2	3	4	5	6
13.	Leaf folder (unidentified)	Folds the tender leaves and feeds higined inside.	September Oct-Nov. Dec.-Aug.	Mild Moderate No incidence	Newiy recorded on cashew crop.
14.	Chelaria Haligramma M.	Feeds on shoot tips, flowers, apple and nuts.	Jan-Feb. March-May June-Dec.	Moderate Severe No incidence	Regularly occurs in flowering season.
15.	Nephopteryx Sp.	Feeds on apple and nuts.	Jan-Feb. March-May June-Dec.	Moderate Severe No incidence	Do
16.	Euproctis Scientillans W.	Defoliation and feeds gregariously on apples and nuts.	Jan-Feb. March-May	Moderate Severe.	Sporadic
17.	Tosoptera odinac Vdg.	Sucks the sap from tender leaves and apples and nuts at peanut state	Feb.-April May-Jan.	Moderate Mild.	Minor
18.	Planococus Sp.	Sucks the sap from panicles and young fruits	April-June July-March	Moderate Mild.	Rare pest.
19.	Flower bedles (unidentified)	Sucks the sap from inflorescence.	Jan.-March	Severe	—
20.	Sthenias grisator	Girdles the stem by forming ring.	Not seasonal	—	Occasional
21.	Unidentified	Damages branches	June-Sept. Oct-May.	Severe Moderate	—

13

Findings of the Study

Introduction

The exact area under cashewnut is very difficult to be estimated owing to unorganised growth in private and forest lands. The accounting procedure differs according to different agencies. For example, there has been an overreporting of cashew area by the Directorate of Economics and Statistics, Bangalore when compared to the estimates provided by the Directorate of Cashewnut Development, Government of India at Cochin. According to the State estimate, the yield rate which was 646 kgs/ha in 1963-64, increased to 846 kgs/ha in 1970-71, and to 1397 kgs/ha in 1974-75. But in the following year, the yield rate declined to 497 kgs/ha and during 1978-79, it stood at 77 kgs/ha.

In Dakshina Kannada, cashewnut is an important crop. In keeping with the World Bank Project to enable higher production and export of cashewnut, a vigorous attempt has been made to convert the wastelands into new cashew plantations. There are over 20 major cashew processing factories employing over 2.0 lakh labourers and with installed capacity of 4.5 lakh tonnes of rawnuts. These factories have to depend largely upon imports from the African countries to realise even half of the installed capacity. Table 13.1 present area and production estimates of cashewnut in the district. The production of rawnuts which was 13,084.4 mt in the base year, increases to 14,276.4 mt in the terminal year, giving a yield rate of 700 kgs/ha in 1979-80 and 491 kgs/ha in 1986-87. When the new plantations would start yielding, it is estimated that the productivity trend will be more than doubled over the preproject average.

Target and Achievement

The area targeted and achieved may be described in respect of the project. Table 13.2 persents the target and achievement from 1980-81 to 1985-86. It is seen that except in the year 1982-83 when the target was overachieved, in other years, there has been a shortfall in the achievement of the targets. During 1980-81, the percentage of achievement was about 90.0 per cent, it increasing to 99.6 per cent in 1981-82, but declining to about 62.0 per cent during 1983-84 and 1984-85. During 1985-86, the achievement rate was 86.0 per cent.

We can look into the talukwise performance in achievement of the targets. The targets and achievement for the respective taluks are presented in Tables 13.3 and 13.4. In these tables, the target and achievement for 1986-87 are also added. The area under cashewnut in 1979-80 was 18,692 ha and by the close of the project period increases to 29091.0 ha in the district. Of the taluks, Bantwal, Puttur, and Sullia had substantial area under cashewnut in the preproject period. After the project, Karkal has also achieved a significant increase in area. Other taluks have also noted marked increase, especially those of Udupi and Kundapur.

During 1980-81 and 1981-82, all the taluks seem to achieve the targets in a very large way except Puttur and Udupi in 1981-82. After 1982-83, there seems to be a notable difference in gross and net targets set out. During 1982-83, the target is over achieved except in the taluks of Karkal and Udupi. During the subsequent year, there has been a shortfall in achievement of the target in almost of the taluks save in Bantwal. In Puttur, Udupi, Karkal and Kandapur, the gap seems to be quite marked. During 1984-85 also, the gap in Karkal, Udupi and also in Kundapur seem to be large. During 1985-86 the performance of achievement was better; even here Udupi and Karkal establish a poor performance. However, during 1986-87, the net target has been overachieved (The figures relate upto September 1986 only). Thus among the different taluks, the percentage of net target achievement has been: Mangalore 86.0, Bantwal 99.8, Puttur 62.5, Sullia 74.7, Belthangady 74.1, Karkal 47.6, Udupi 39.2 and Kundapur 69.6 per cent respectively. Thus, it is seem that Udupi and Karkal have registered poor target achievement when compared to other taluks.

But, for the district as a whole, the net target has been more or less achieved except in the taluks of Udupi and Karkal as mentioned above. In this district, there is a vast scope for expanding ca shew cultivation in wastelands. According to Table 13.5, an estimated area of 61,395.56 ha is available as wasteland, of which the targeted area achieved (10,399 ha) is included. When the area of wasteland and wasteland brought under cashew cultivation is considered, satisfactory coverage is noticed in Mangalore, Bantwal, Kundapur and to a certain extent in Sullia. In Belthangady, the achievement is more than the targeted limit. So, it is only in the case of Udupi and Karkal that the wasteland brought under cashew cultivation is only about 7.2 and 8.5 per cent respectively of the available wasteland. This factor is solely responsible for poor performance in the above taluks. Thus, there is ample scope to utilise the available wasteland in Udupi and Karkal for cashew cultivation. Also, in Puttur, the encouragement of rubber plantations has impaired the growth of cashew crop. It is evident that unfavourable conditions in utilising the wastelands was a major factor which calls for high capital investment for land development and planting of cashewnut seedlings.

Subsidy and Beneficiaries

Nearly 40.0 per cent of the area covered under the new plantations has been earmarked for subsidy accounting as shown in Table 13.6. Over the years, the area has been increasing from 122.62 ha in 1980-81 to 835.99 ha and 698.50 ha in 1984-85 and 1985-86 respectively. The talukwise performance shows that Bantwal, Karkal, Udupi and Kundapur have fased well in subsidy accounting in absolute figures. However, if the subsidy accounting area as a percentage of new area brought under cashew cultivation is considered, it is seem that the achievement under this comes to about 40.0 per cent in almost all the cases as in the case of district average.

During 1980-81, a total number of 263 beneficiaries with an area of 336 ha was covered. The number increased to 1173 in 1981-82 and the area to 1458 ha. Whereas the coverage fo farmers was higher in Karkal, Udupi and Kundapur in

1980-82, excepting Mangalore and Sullia which reported lower coverage even during 1982-83. The number of beneficiaries increased to 1929 in 1983-84, to 1888 in 1984-85 and 2180 in 1985-86. The coverage of farmers in the taluks of Udupi, Kundapur and Karkal seem to be quite substantial when compared to other taluks. It is seem that over 8768 cashew cultivators have been benefited by the Project with an area of over 9309 ha.

Table 13.7 describes the number of beneficiaries who have received financial assistance sourcewise during the first four years of the Project. The primary land Development Banks gave better than the Commercial Banks in providing assistance to the cashewnut cultivators. Of the total beneficiaries, over 99.0 per cent received assistance from the PLDBs. The area covered also was very substantial in their case. This suggests that the Commercial Banks have to come forward vigorously in cashewnut development of the district. The commercial banks have some noticeable coverage in Udupi, followed by Karkal.

Under the Project, training to farmers in improved methods of cashew cultivation (for example, in top working and vegetative propagation) have been started since 1985-86 and about 1200 farmers from the different taluks are being trained now. The new plantings would come to bearing stage within 4 to 5 years and afterwards the yield rate increase would be substantial. In the tenth year, the increase is expected to be about 750 kgs per ha. At full project development, about 2100 tonnes of cashew would be produced annually. By the 10th year, the addition to existing production would be approximately 700 tonnes. At the fourth year the average yield of new plantations would be about 40 kgs/ha and in the 10th year it could increase upto 750 kgs/ha. When the projects is evaluated, it seems that there would be a surplus after meeting all expenses from the 10th to the 40th year of new plantations.

TABLE 13.1 : **Area and production of cashew in Dakshina Kannada**
(area in ha and production in mt)

Year	*Area*	*Production*
1979-80	18,692.0	13,084.4
1980-81	18,995.0	13,084.4
1981-82	20,413.8	13,084.0
1982-83	22,009.2	13,094.4
1983-84	24,098.2	13,084.4
1984-85	26,194.3	13,114.7
1985-86	27,919.4	13,406.4
1986-87	29,091.1	14,276.4

TABLE 13.2 : **Progress under Multi-State Cashew Project in D.K.**
(in ha)

Year	*Target*	*Achievement*
1980-81	338.0	303.0
1981-82	1424.0	1418.8
1982-83	1450.0	1595.4
1983-84	3392.0	2089.0
1984-85	3300.0	2096.1
1985-86	2000.0	1725.1
Total	12584.0	9227.4

Source : Deputy Director of Horticulture (Cashews), D.K. Mangalore.

TABLE 13.3 : **Yearwise and Talukwise Targets during the Project period from 1980-81 to 1986-87 (in ha).**

Sl. No.	*Taluk*	*Pre-Project period area 1979-80*	*1980-81*	*1981-82*	*1982-83*	*1983-84*	*1984-85*	*1985-86*	*1986-87*	*Total*
1.	Mangalore	1464	10	110	100	201	75	60	50	606
2.	Bantwal	3358	30	150	202	354	275	200	150	1361
3.	Puttur	2700	15	318	218	440	300	200	55	1566
4.	Sullia	5686	40	110	82	190	150	120	75	767
5.	Belthangady	2075	30	106	200	410	250	200	150	1346
6.	Karkal	926	90	318	584	954	1660	740	450	4796
7.	Udupi	967	60	318	642	1032	1660	740	450	4902
8.	Kundapura	1516	60	170	400	726	550	420	200	2525
	Pre-Project *TARGET* : figures are Gross :	18692	338	1600	2428	4307	4800	2680	1600	17753
	as per the Deptl. figures 1979-80. *Net* :		338	1424	1450	3392	3300	2000	1000	10904

TABLE 13.4 : **Yearwise and Talukwise Achievements during the Project period from 1980-81 to 1986-87 (in ha)**

Sl. No.	Taluk	Pre Project Period area 1979-80	1980-81	1981-82	1982-83	1983-84	1984-85	1985-86	1986-87	Total
1.	Mangalore	1464	10-00	91-00	66-80	104-20	109-20	63-60	76-60	521-40
2.	Bantwal	3358	29-60	173-00	192-20	376-30	277-50	213-00	96-10	1357-90
3.	Puttur	2700	12-40	181-90	163-20	243-60	215-60	121-60	40-80	979-00
4.	Sullia	5686	36-40	177-80	59-40	88-00	141-20	101-20	69-20	573-20
5.	Belthangady	2075	26-20	170-60	122-20	178-80	217-20	200-00	82-80	997-80
6.	Karkal	926	89-60	306-60	329-00	364-60	403-40	449-20	344-00	2286-40
7.	Udupi	967	52-00	179-40	330-00	377-60	445-80	227-30	313-00	1925-00
8.	Kundapur	1516	46-80	238-40	332-90	355-90	286-20	349-20	149-20	1758-30
	Pre-Project figures are as per the Deptl. figures 1979-80	Gross : 18692	808-00	1418-40	1595-40	2089-00	2096-10	1725-10	1171-70	10399-10

TABLE 13.5 : **Wasteland Utilisation in Dakshina Kannada (in ha)**

Taluk	*Pre-project*	*Covered during 1980-81 to 1986-87*
Mangalore	696.12	521.40
Bantwal	1,102-40	1,353.90*
Puttua	2,800-20	979.00
Sullia	926.44	573.20
Belthangady	365.60	997.80
Udupi	26,801.54	1,925.10
Kundapura	1,922.41	1,758.30
Karkal	26,780.85	1,286,40
District	61,395.56	10,399.10

*includes nanja lands highlying rainfed paddy field.

Source : District Horticultural Office (Cashew), Mangalore.

TABLE 13.6 : **Area Sanctioned for Subsidy accounting (ha)**

Taluk	*1980-81*	*1981-82*	*1982-83*	*1983-84*	*1984-85*	*1985-86*	*1986-87*	*Total*
Mangalore	4.05	36.83	26.11	40.47	44.52	25.90	30.76	209.64
Bantwal	10.12	70.01	77.70	144.07	109.27	85.79	39.26	536.22
Puttur	2.83	73.65	65.97	94.70	87.41	33.59	16.19	374.34
Sullia	8.09	31.57	23.88	33.59	57.06	48.16	27.92	230.27
Belthangady	12.55	69.20	49.37	61.11	78.11	86.60	42.09	399.03
Karkal	25.50	124.24	133.14	150.55	163.10	152.17	131.12	876.82
Udupi	16.19	72.44	177.36	144.88	176.85	124.65	126.65	779.04
Kundapur	14.57	136.79	134.76	140.84	119.39	141.64	60.71	748.70
Total	122.62	574.26	629.30	811.82	835.29	698.29	474.30	4150.00

Source: India Cashew Project, Mangalore.

TABLE 13.7 : No. of beneficiaries and area (ha) by Bank-Sources

Taluk	No. of beneficiaries		Area (ha)	
	PLDB	Com. Banks	PLDB	Com. Banks
Mangalore	246	15	319.7	20.4
Bantwal	859	8	1,000.4	7.5
Puttur	621	9	710.9	13.8
Sullia	308	6	382.4	7.1
Belthangady	695	1	689.4	1.8
Karkal	1,027	54	1,298.9	63.5
Udupi	606	185	848.6	187.9
Kundapur	1,002	19	1,263.3	26.5
District	5,364	297	6,513.6	328.5

Source: District Horticulture Office (Cashew), Mangalore.

14

Improved Cashew Cultivation

An attempt is made to evaluate the performance of improved cashew cultivation introduced by the World Bank Project in D.K. since 1980-81. Of the 8 taluks in the district, 6 have been selected for evaluating the impact of the project on cashew cultivation. Since 1980-81, farmers have been getting subsidy of about Rs. 900 per ha of the new cashew plantations. A sample of 150 units as presented in Table 14.1 have been studied in Bantwal, Belthangady, Karkal, Mangalore, Putter and Udupi. The farmers were mostly small holders with an average holding of 1.9 ha for the entire sample. Along with the farm units which had the benefit of subsidy, some of the farms which did not enjoy the benefit were also considered. Since it was a question of bringing in new land (especially wasteland) into cashew cultivation only these farmers who had the capacity of acquiring such land and came forward for cashew cultivation were selected for the subsidy programme. Strictly speaking, the yield obtained in new lands could not be compared with that in the established old lands due to differences in soil fertility, etc.

A peculiar feature of the programme was that in the plantations, only the improved Ullal cashew verieties were propagated. Quality seedlings were distributed to the farmers with technical advice relating to fertilizer application, plant protection measures and other cultural practices. Every year the subsidy was given. In the latter years, commercial bank in the district also showed interest in extending financial assistance to the cashew growers.

For the non-subsidy farmers, the average landholding was 1.75 ha, slightly lower than the subsidy cases. The distribution according to the taluks is presented in the same Table. Except in Belthangady and Mangalore where the average landholding was more than two ha, in other places, it varied from 1.25 ha to 2.0 ha.

Table 14.2 describes the cropping pattern in the subsidy and non-subsidy cases in the sample regions. In Bantwal, paddy accounts for 16.5 per cent, coconut 4.2 per cent, arecanut 2 per cent and cashewnut 30 per cent of the total cropped area. It would be seen that the major crops in the region denote different patterns of cultivation according to the landscape and water regime. Paddy followed by arecanut and coconut demand high water regime. Cashew on the other hand is a drought-tolerent crop and grows well even when the rainfall is scanty.

The proportion under the crops as presented in the Table shows that in subsidy case, the average area under paddy is 18.3 per cent, that of cashewnut 26.5 per cent, while coconut and arecanut share 5.2 and 5.1 per cent respectively. In Mangalore region, paddy share is 26.6 per cent and in this case, the share of cashew is much below the sample average. Coconut has more than the average share in Karkal and arecanut in Puttur.

In the non-subsidy cases, the proportion of area under paddy in slightly higher at 21.8 per cent, it being still higher in Kerkal, Mangalore and Udupi regions. In coconut and arecanut also, the non-subsidy group has a higher proportion. Naturally, the area under cashew would be lower. In this case, it is 23.0 per eent, it being higher in only one region—Bantawl.

Distribution of Trees by Age and Yield

Among the sample households, the distribution of cashew trees according to three different age groups may be seen. The groupings attempted here are trees of less than 3 years, between 3-5 years, between 5-10 years, between 10-20 years and above 20 years. It is hardly seen that area under cashew is not reported by the farmers till they come of bearing age. Usually, that particular stage happens to be after the 5th year. It has been found out that even in the case of trees of 3-5 year range, some yield has been recorded in the sample households. One or two trees may yield by the 3rd year end itself.

If the area under cashew of less than 3 years is excluded from the total cashew area, net area will decline substantially. The range of proportion under 3 year old trees is between 6-12.7 per cent in the case of local varieties. Among the subsidy households, care has been taken to include the bulk of the cases which have reported cultivation of improved varieties in the initial years of introduction, viz., 1960-81 and 1981-82. Most of the farmers got subsidy in the subsequent years also and hence reporting of below 3 years old trees is also very common. But its proportion does not increase beyond 23.6 per cent and is as low as 4.8 per cent in some cases. No subsidy case growing new plantation beyond the 5th year was normally encountered in the samples. In one or two instances, the improved varieties have been grown since 7 or 8 years due to the progressive nature of the farmers and their frequent contact with the cashew research stations. Table 14.3 presents the age-wise distribution of cashew trees in the sample region.

The Table describes the distribution of cashew trees according to age in respect of the subsidy samples for both local and improved varieties. Less than 3 year old trees account for 6 per cent for locals in Bantwal. The proportion under the subsequent groups (3-5 years—II, 5-10 years—III, 10-20 years—IV and above 20 years—V) is 4-5, 47.0, 20.1 and 22.4 per cent respectively. In Belthangady sample, the proportion under the groups are 7.8, 3.7, 56.1, 18.3 and 41.1 per cent. In Karkal, the proportion under the first group is slightly higher—12.7 per cent and in the subsequent groups, the proportion is 8.1, 60.1, 23.5, and 4.6 respectively. In the Mangalore sample, the share under the 5 groups is 9.2, 8.0, 53.1, 18.1, and 11.6 respectively. In Puttur and Udupi, II group was absent and the proportion under the other groups is 8.2, 26.0, 56.0 and 9.8 per cent in Puttur and in Udupi 6.3, 36.0, 50.0 and 7.7 per cent respectively.

It is seen that in almost all the regions, the distribution under 5-10 years and 10-20 years has been marked. Though the subsidy farmers grow local varieties also, the proportion under the same is less than 10 per cent in almost all the cases. In Puttur and Udupi regions, tree population of over 10 years of age is maximum among the samples. Though local cashew trees can live for over 30-40 years of age, it is doubtful whether trees over 30 years will yield the optimal yield rates. Not only that, the old trees are prone to attack by various pests and diseases and hence their yielding capacity would be lower. Moreover the rooting system of the trees may not protrude deep into the soil during this period and coupled with other factors, flower and fruit formation, fertilizer uptake capacity, etc., may be impaired. The damage done by the stem borer is more pronounced in the later years of the cashew tree. When that particular pest attacks the tree, there is bound to the its collapse in a year or two. Judged from the above, it could be hypothesized that trees of over 20 years are most suitable for top-working.

It is in respect of the improved varieties that the analysis is concerned most. The over 5 years trees are found to a small extent in Mangalore and Udupi region only. In others, the proportion under the II group (3-5 years) is over 73 per cent. Hence the proportion under the I group ranges from as low as 4.8 per cent in Udupi to 23.6 per cent in Bantwal.

It may interesting to note the distribution of the trees in the case of non-subsidy units also (Table 14.4). Hence though some of the non-subsidy farms in Bantwal and Mangalore grow improved varieties in a small area, the bulk of coverage is under local varieties only. In the I group, the distribution of trees ranges from 3.6 in Belthangady to 102.2 per cent in Puttur. Under the II group, the range is from 9.8 per cent (Mangalore) to 8.2 per cent (Belthangady). Karkal has only 18.4 per cent trees under the III group, Udupi reporting the maximum of 32.1 per cent. It could be seen that the proportion of trees in the last group is higher than that in the I group. In the IV group, the range is from 26.0 per cent is Belthangady 42.2 per cent in Mangalore. The last group, unlike in subsidy cases reports a higher proportion in the non-subsidy cases. It is least in Udupi (12.5 per cent) and increases upto 31.2 per cent in Karkal. When both the subsidy and non-subsidy cases are compared for this group, it seems that in the former, a certain proportion of old trees have been cut. It could be that the diseased trees are cut so that new plants in the area are not affected by that particular disease or pest.

An attempt is made to evaluate on the proportion of yielding to non-yielding trees. The analysis pertains to only two groups— less than 5 years and over 5 years. Under these 2 groups, the proportion of yielding trees for both local and improved varieties is presented in Table 14.5.

In Bantwal, only 6.3 per cent of local trees yield under less than 5 years group. The proportion of the II group is 69.7 per cent. For the non-subsidy cases the proportion works out to 7.0 and 68.5 per cent respectively. In Belthangady, the non-subsidy cases have slightly lower proportion of yielding trees of all the samples. Belthangady has the highest yielding proportion (88.2 per cent for over 5 years old trees). In Karkal, the non-subsidy samples have an edge over the subsidy samples in yielding proportion (12.0 per cent) under I group. For the II group, there is no variation at all, the proportion being 68.3 per cent in both the samples. In Mangalore, while a slightly higher proportion is reported for non-subsidy locals in the I group, it is the subsidy samples that report a higher proportion of yielding trees under the II (group 82.3 per cent). In Puttur and Udupi areas, the proportion under the I group is very low, the non-subsidy cases enjoying a slightly advantageous position except in the subsidy II group.

For the improved varieties, very few instances have been noted and the yielding proportion for the I group is 6.6 per cent in Bantwal and 8.5 per cent in Mangalore. It is to be noted that cultivation of improved varieties has been spontaneous on the farmers' side even before the inception of the World Bank and NABARD Projects.

In the subsidy improved cases, the proportion of yielding trees in the I group ranges from a mere 6.2 per cent in Puttur to 13.8 in Karkal and 16.2 per cent in Belthangady. For other regions, variations are marked within this range. Only Mangalore and Udupi report a few cases of trees of over 5 years old and there, the yielding proportion is over 91.5 per cent.

When the performance of the two varieties is compared in all the places, the yielding proportion in improved variety seems to be higher thah that in the local varieties. The yielding proportion gap is big in Karkal and Mangalore where the proportions are 13.8 and 8.5 in Karkal and 10.3 and 4.5 per cent in Mangalore respectively.

Cashewnut Productivity in regions

An attempt is made to evaluate the yield capacity of cashewnut in the subsidy and non-subsidy households for different age groups of trees. Table 14.6 per cents the yield variation for subsidy cases only. The yield rate for local varieties under the three year old trees has been 0.5 kg per tree in Mangalore and Belthangady, 0.1 kg per tree in Puttur. The average figure shows that the yield rate of local varieties is 0.23 kg per tree and for improved varieties it is slightly higher at 0.3 kg per tree. Taking an average number of 200 trees per ha, the yield rate for local under 3 year old group would be 46 kg per ha and for improved varieties 60 kg per ha.

In the II group (3.5 years), the average productivity of local varieties is 1.9 kg per tree, working out to 380 kg per ha. Across the region, the yield level is up to 2.5 kg per tree in Puttur and Belthangady, 1.6 kg per Bantwal and Karkal.

In the III group (5-10 years old trees) productivity figures for improved varieties were not available save in Mangalore and Udupi. In these two places, the average productivity has been 4 kg per tree as against the local average of 3.7 kg per tree for all the regions. Among the different regions, the yield variation is between 2.5 kg and 2.8 kg in Mangalore and Bantwal to 3.5 kg in Udupi, 3.7 kg to in Belthangady and 4.5 kg per tree in Puttur.

For the local varieties, the average yield rate has been 5.7 kg for the IV group and 6.8 kg for the V group. Not much difference is noticed in the groups except in areas like in Karkal in and Mangalore there is some substantial increase in yield in the V over the IV group. In Puttur for example, trees in the last group yield a lower level than the preceding group.

From Table 14.6 it is seen that the average yield rate for local varieties increases from the base level of 46.0 kg per ha to 1136 kg per ha for the very mature trees. In the case of improved varieties, yield increase has been from 0.3 to 4.0 kg per tree in the III group. The yield increase under local verieties between the I and III group is more or less similar to that under the improved varieties, though the yield potentiality of the latter seems to be quite high.

Here, it could be seen that the very old trees of over 25 to

30 years yield less than the optimal level. Strictly speaking, very little economics could be visualized for unthrifty trees because whether they yield or not the farmers may not consider them with preference. Such trees may not command application of manures and fertilizers and other cultural practices. Though in some cases the per tree yield is more than 10 kg, equally low yields have also been reported.

Table 14.7 describes the yield differentials for the non-subsidy cases. The average yield rate under the I group is 0.22 kg per tree as against the subsidy average of 0.23 kg per tree. Except in Belthangady and Mangalore regions, per tree yield is less than 0.2 kg in other places. For the II group, the average yield is slightly higher by about 4 per cent than the subsidy cases (1.97 as against 1.9 kg). Again, Mangalore and Belthangady along with Karkal have more than the average yield. In the III group, however, the average yield rate is slightly less than the subsidy average. In the non-subsidy case, per ha yield rate has been 720 kg whereas in the subsidy cases, it has been 740 kg.

In the IV and the subsequent groups the average non-subsidy yield rate has been higher than that under the subsidy cases—1274 per ha against 1140 in the IV group and 1526 as against 1360 kg per ha in the last group. Unlike the subsidy cases, it is difficult to establish the falling yield trend after the IV group level. The number of such cases encountered have been few owing to the fact that the very proportion of population under the last group has been very low.

Organic manure and Fertilizer application

The nutrient removal in a 30 year old cashew tree has been 2.8 kg N, 0.8 kg P_5O_2 and 1.3 kg K_2O. Therefore, it is essential that depending on soil fertility, organic manure and synthetic fertilizers are applied to the cashew trees of all age groups. The common practice in the region is that the farmers would apply organic manure consisting of cattledung, farmyard manure, green leaves and ash especially in nut-bearing stage of the tree. With the introduciion of the cashew development programme, application of macro nutrients to improved cashew varieties has become very essential. It is also observed that flower and fruit formation would improve with higher dosage of nutrients. The fertilizer could be applied in a single dose or in two split doses.

Table 14.8 describes manure and fertilizer application per tree for subsidy and non-subsidy cases. The average quantity of organic manure applied to local varieties in subsidy case is about 1.5 kg per tree. In Karkal and Mangalore, the dosage is more than the average. Even for local varieties chemical fertilizer have been applied, the average for the district being about 0.17 kg per year. For the improved varieties, the dosage of both organic manure and chemical fertilizer is higher—1.63 kg manure and 0.68 kg fertilizer per tree. In the non-subsidy cases, not all the farmers apply chemical fertilizers to local varieties. In places where the application is being made (Belthangady and Mangalore), the average dosage works out to 0.15 kg per tree. Some of the farmers grow improved varieties of their own and apply on an average 0.2 kg per tree per year except in the regions of Karkal and Puttur.

Productivity of major crops in the regions

It may be profitable to view the performance of some major crops grown in the region along with cashewnut. The major crops are, however, of irrigated nature and have little to do with cashew yield or cultivation. There has been a diversion of investment, manure and fertilizer to cashewnut might influence the yield rates of the major crops which hitherto, directly or indirectly got the benefit of cashewnut manure and fertilizer share. Table 14.9 presents the productivity of paddy, arecanut and coconut for subsidy and non-subsidy cases.

It could be seen that the average yield of paddy for the entire sample is 3760 kg per ha, whereas for non-subsidy cases it is 3728 kg. Among the regions, Puttur, and Udupi seem to have a higher average yield rates. The yield rate in non-subsidy cases is higher in Bantwal, Belthangady and Udupi by over 1.2, 19.0 and 1.6 per cent respectively. Subsidy yield rates are higher by 4.8, 1.0 and 13.5 per cent in Karkal, Mangalore and Puttur respectively. But, on the whole, the yield differential of just 1.0 per cent between the subsidy and non-subsidy cases points out the evenness.

In the case of arecanut, the average yield in subsidy cases is 1176 kgs per ha and higher in the non-subsidy cases at 1369 kgs. This increased is especially visible in almost all the non-subsidy areas. The increase is as high as 28.0 per cent in Mangalore for the average increase of 16.3 per cent for all the cases. In Belthangady and Karkal, non-subsidy arecanut cultivation has not been covered.

For coconut, the trend is similar to that of paddy, the non-subsidy yield rate being lower by 3.3 per cent. In Belthangady, Karkal and Udupi, the yield rate in the subsidy cases is higher. Bantwal non-subsidy cases report yield increase or 13.8 per cent and in Puttur records a marginal increase. In Udupi, the increase is by 10 8 per cent.

From the above, it seems that yield pattern between paddy, arecanut and coconut has no trend and regional specifications, count much in yield differentials.

Marketing of Cashew

Raw cashewnuts are harvested during March-Apıil and the first pluck is immediately disposed of. The second pluck is being kept for 4-6 weeks depending on the market trend and sold to local merchants and to taluk headquarters commission agents. Usually the nuts are collected after the fruit is ripe but now-a-days, theft, rising prices, etc., induce the farmers to collect tender nuts and market the same. This tendency reduces not only to yield rate but also the quality of the nuts. Though it has been observed that the proportion of tender nuts collected is 15.0 in subsidy cases and 13.2 per cent in non-subsidy cases, it may be presumed that the percentage may be quite high in recent years (Table 14.10). Harvesting of nuts at tender stages affects storage quality also. Due to increasing prices, the small farmers, especially to get some cash, dispose of the produce. They could have got some advance also from the merchants and may be obliged to market the produce of lower prices to those merchants.

The percentage of tender nuts plucked is very high in Mangalore, Puttur and Udupi for both subsidy and non-subsidy cases. In other places, though the proportion is lower than 12.0 per cent, the tendency is that every year the proportion would increase. Even 15 to 30 days before fruit ripening stage, the nuts are collected. This process also denies collection of higher quantity of quality cashew apples.

It has been found that almost all the farmers have started practising pruning of old and unthrifty trees as and when necessary. The diseased trees are cut and are either used as fuel or sold. Cashew tree is also used in boat construction, etc. As a matter of fact, the unthrifty cashew trees which are felled during the rainy season account for the major source of fuel-wood in the regions. It has been observed that even yielding trees are cut to be sold for boat construction, pulp-making, etc.

Table 14.11 presents marketed grades of raw cashewnut and prices in the different regions for both subsidy and non-subsidy cases. In all the places, only three major grades are marketed based on the colour and size of the rawnuts. In Bantwal, I grade nuts account for 23.0 per cent of the total marketed quantity, while the II grade and III grades account for 51.0 and 26.0 per cent respectively for the subsidy cases. For the non-subsidy cases, the proportions are 25.0, 60.0 and 15.0 per cent respectively for the three grades. The price for kg of I grade is Rs. 11.5 and Rs. 11.3, whereas for the other grades, it is Rs. 10.6 and Rs. 10.5 and Rs. 8.5 and Rs. 9.8 respectively for the subsidy and non-subsidy cases.

In Belthangady, the proportion under the three grades is 13.0, 54.0 and 23.0 per cent under subsidy and the 14.0, 63.4 and 22.6 per cent under non-subsidy cases. The prices are slightly higher than in Bantwal, but the non-subsidy cases present a slightly advantageous price level. In Karkal, the proportion under I grade is as low as 9.6 per cent and the proportion under III grade is as high as 38.0 per cent for subsidy cases. For non-subsidy cases, the proportion under I grade is only 11.0 per cent. The price variation is almost nil between the two cases. In Mangalore, the price level is lightly lower and in the non-subsidy cases, the proportion under III grade is low (similar to Bantwal and Puttur). The price variation is quite attractive in Puttur. Hence there is a lot of difference between subsidy and non-subsidy cases in better marketing of quality cashew. The proportion under I grade is only 13.5 per cent in subsidy cases, but as high as 55.0 per cent in non-subsidy cases. The proportion under III grade is 47.5 per cent for subsidy cases, and 10.0 per cent for non-subsidy cases. In Udupi also, the proportion under I grade in non-subsidy cases is higher than in the sobsidy cases (31.0 as against 17.0 per cent) and the difference in III grade is not much owing to the higher proportion under II grade.

For all the regions, it is seem marketed quantity is high, almost half in second quality. If farmers could practice ripe fruit collection, it is sure the quantity marketed in I grade could be increased. From Table 14.10 it is evident that the non-subsidy cases reporting a lower proportion of tender nut plucking, has much to do with higher proportion of I grade nuts marketed in places like Puttur and Udupi.

The quantity of cashew apples harvested comes to about 3700 kg per ha in both the subsidy and non-subsidy cases. About 1 to 2 per cent is consumed by human beings, about 6-8 per cent by cattle and the rest disposed off as manure or waste. In some places distillation of liquor from cashew apple is also practised illegally.

New plantations adhere to there commended spacing of 24 × 24 feet for all the trees. The subsidy is uniform at Rs. 900 per ha in all the places. The loan amount is Rs. 3900 per ha in Bantwal, it being the highest in the sample. It is Rs. 3710 and Rs. 3720 respectively in Puttur and Belthangady, followed by Rs. 3680 in Karkal and Udupi and Mangalore reports the lowest loan amount of Rs. 3510. In some cases, the farmers have obtained loan amount for more than one year for new plantations as well as for improvement of the existing plantations (especially for applition of manures and fertilizers).

TABLE 14.1 : **No. of sample farmers and area (ha)**

Region	*Subsidy*		*Non-subsidy*	
	Nos.	*Area*	*Nos.*	*Area*
Bantwal	16	20.0	16	18.6
Belthangady	30	60.5	18	26.3
Karkal	26	46.8	16	19.8
Mangalore	25	80.2	15	58.3
Puttur	28	43.2	18	21.0
Udupi	25	35.0	17	23.5
Total	150	285.7	100	174.5
Per H/H area	—	1.90	—	1.75

TABLE 14.2 : **Cropping pattern in the Subsidy (S) and Non-subsidy (NS) Samples (per cent)**

Region	*Paddy*		*Coconut*		*Arecanut*		*Cashewnut*	
	S	*NS*	*S*	*NS*	*S*	*NS*	*S*	*NS*
Bantwal	16.5	12.9	4.2	5.0	2.0	1.0	30.0	28.0
Belthangady	16.2	16.5	5.8	9.6	4.0	4.0	24.0	20.0
Karkal	14.8	30.0	8.5	3.0	2.3	0.2	25.0	21.0
Mangalore	26.6	32.0	2.5	8.0	2.6	7.5	19.0	16.5
Puttur	10.0	8.0	5.2	6.0	18.5	28.8	26.5	23.1
Udupi	18.6	25.0	3.5	2.0	1.0	6.0	26.0	22.0
Total	18.3	21.8	5.2	5.7	5.1	7.9	26.5	23.0

TABLE 14.3 : **Distribution of Cashew trees in subsidy units (per cent)**

Region	*less than 3 years*		*3-5 years*		*5-10 years*		*10-20 years*		*above 20 years*	
	L	*I*	*L*	*I*	*L*	*I*	*L*	*I*	*L*	*I*
1	*2*	*3*	*4*	*5*	*6*	*7*	*8*	*9*	*10*	*11*
Bantwal	6.3	23.6	4.5	76.5	47.0	—	20.1	—	22.4	—
Belthangady	7.8	20.5	3.7	79.5	56.1	—	18.3	—	14.1	—
Karkal	12.7	23.1	8.3	76.9	60.1	—	23.5	—	4.6	—
Mangalore	9.2	18.3	8.0	79.4	53.1	2.3	18.1	—	11.6	—
Puttur	8.2	22.0	—	78.0	26.0	—	56.0	—	9.8	—
Udupi	6.3	4.8	—	86.6	36.0	8.6	50.0	—	7.7	—

L=Local;
I=Improved.

TABLE 14.4 : **Distribution of Cashew trees in non-subsidy-units (per cent)**

Region	*Less than 3 years*	*3-5 years*	*5-10 years*	*10-20 years*	*about 20 years*
Bantwal	4.5	12.3	20.3	34.5	28.2
Belthangady	3.6	18.2	31.2	26.0	21.0
Karkal	6.8	13.4	18.4	30.2	31.2
Mangalore	8.3	9.8	21.5	42.2	18.2
Puttur	10.2	12.1	28.3	29.3	20.1
Udupi	7.3	12.5	32.1	35.6	12.5

TABLE 14.5 : **Proportion of yielding trees**

Region		*Subsidy*		*Non-subsidy*	
		Local	*Improved*	*Local*	*Improved*
Bantwal	1.	6.3	9.8	7.0	6.6
	2.	69.7	—	68.5	—
Belthangady	1.	14.0	16.2	12.5	—
	2.	88.2	—	80.2	—
Karkal	1.	8.5	13.8	12.0	—
	2.	68.3	—	68.3	—
Mangalore	1.	4.5	10.3	6.0	8.5
	2.	82.3	100.0	78.0	—
Puttur	1.	3.2	6.2	4.8	—
	2.	80.1	—	81.2	—
Udupi	1.	6.0	8.8	7.0	—
	2.	83.5	90.5	81.6	—

1.=less than 5 years;
2.=above 5 years.

TABLE 14.6 : **Yield level of Cashew trees (kg/trees)—Subsidy**

Region		*Less than 3 years*	*3-5 years*	*5-10 years*	*10-20 years*	*above 20 years*
Bantwal	L	0.05	1.5	2.8	4.9	5.1
	I	0.2	2.0	—	—	—
Belthangady	L	0.5	2.5	3.8	6.5	6.3
	I	0.3	2.5	—	—	—
Karkal	L	0.1	1.5	5.0	6.1	12.0
	I	0.5	2.8	—	—	—
Mangalore	L	0.5	1.6	2.5	5.3	6.5
	I	0.1	2.0	3.9	=	—
Puttur	L	0.1	2.5	4.5	5.5	4.6
	I	0.1	3.3	—	—	—
Udupi	L	0.1	1.6	3.5	5.8	6.0
	I	0.2	2.0	4.1	—	—
Average	L	0.23	1.9	3.7	5.7	6.8
	I	0.30	2.4	4.0	—	—
Yield/ha	L	46	380	740	1140	1360
	I	60	480	800	—	—

L=Local ; I=Improved.

TABLE 14.7 : **Yield level of Cashew trees (kg/trees)—non-subsidy**

Regoin	*Less than 3 years*	*3-5 years*	*5-10 years*	*10-20 years*	*above 20 years*
Bantwal	0.1	1.7	2.8	5.2	7.1
Belthangady	0.5	2.0	2.5	6.0	6.8
Karkal	0.2	2.0	4.8	9.0	11.3
Mangalore	0.3	3.1	5.3	8.0	6.5
Puttur	0.1	1.5	2.9	5.1	7.1
Udupi	0.1	1.5	3.5	4.9	7.0
Average	0.22	1.97	3.63	6.37	7.63
Yield/ha	44	394	726	1274	1526

TABLE 14.8 : **Manure and fertilizer application (kg/tree)**

Region	*Subsidy*				*Non-subsidy*			
	Local		*Improved*		*Local*		*Improved*	
	O	*F*	*O*	*F*	*O*	*F*	*O*	*F*
Bantwal	0.6	0.3	1.0	0.6	1.0	—	1.0	0.3
Belthangady	0.8	0.2	1.0	0.7	0.5	0.1	1.0	0.2
Karkal	2.6	0.1	2.0	0.7	1.6	—	1.3	—
Mangalore	2.1	0.2	2.1	0.8	1.8	0.2	1.5	0.2
Puttur	1.3	0.1	1.5	0.8	1.0	—	1.2	—
Udupi	1.6	0.1	2.2	0.5	1.8	—	1.8	0.2
Average	1.5	0.17	1.63	0.68	1.3	0.15	1.3	0.2

O=Organic manure including dung, green manure and ash.
F=Chemical fertilizers including urea, complex, superphosphate, etc, (NPK).

TABLE 14.9 : **Productivity of major Crops (kg/ha)**

Region	*Paddy*		*Arecanut*		*Coconut*	
	S	*NS*	*S*	*NS*	*S*	*NS*
Bantwal	3459	3504	976	1062	1052	12355
Belthangady	3244	3860	1087	—	9513	9236
Karkal	3820	3642	1237	—	11787	9039
Mangalore	2856	2659	1707	2189	12768	10848
Puttur	5127	4509	998	1065	8923	9091
Udupi	4280	4346	1112	1236	11416	12654
Average	3760	3728	1176	1369	10786	10440

S=Subsidy ;
NS=Non-subsidy.

TABLE 14.10 : **Some cultural practices adoption.**

Region	*% of tender nuts plucked*		*% of pruning adopted*	
	S	*SN*	*S*	*NS*
Bantwal	11.5	11.8	52.5	46.0
Belthangady	9.5	11.0	48.0	38.5
Karkal	8.5	9.8	45.1	38.0
Mangalore	16.5	17.3	50.0	40.6
Puttur	21.0	15.0	50.0	35.0
Udupi	23.5	14.9	40.5	30.1
Average	15.0	13.2	47.2	37.7

S=Subsidy ;
NS=Non-subsidy.

TABLE 14.11 : **Marketed grades of raw cashewnut**

Region		*I Grade % of Qty,*	*Price*	*II Grade % of Qty.*	*Price*	*III Grade % of Qty.*	*Price*
Bantwal	S	23.0	11.5	51.0	10.6	26.0	8.5
	NS	25.0	11.3	60.0	10.5	15.0	9.8
Belthangady	S	23.0	12.3	54.0	11.3	23.0	10.5
	NS	14.0	12.5	63.4	11.8	22.6	11.1
Karkal	S	9.6	12.5	52.5	11.9	38.0	11.0
	NS	11.0	12.3	59.3	11.9	29.7	11.3
Mangalore	S	24.8	11.9	50.7	11.1	25.3	10.3
	NS	20.0	11.7	62.0	10.9	18.0	10.1
Puttur	S	13.5	13.0	40.0	12.3	47.5	11.2
	NS	55.0	12.7	35.0	11.9	10.0	11.2
Udupi	S	17.0	12.8	50.0	12.1	33.0	11.4
	NS	32.0	12.4	40.5	11.8	27.5	11.1

S=Subsidy ;
NS=Non-subsidy.

15

Cashew Industry

Cashew cultivation was started for the first time in India in Mangalore in the second decade of this century. Mangalore is thus the birth place of this industry. Thereafter, the industry spread to other parts of the country. Now, nearly 80 per cent of this industry in India is centred in and around Quilon in the Kerala State and Dakshina Kannada accounts for about 12.0 per cent (on the basis of the exports made by the factories). The produce of the district is not sufficient to meet the requirements of the cashewnut factories. Hence large quantities of raw cashewnut are being imported from Africa. Some quantities are also obtained from the Uttara Kannada district and also from Kerala. The raw cashewnuts grown in India and made available to the cashew factories constitute harolyp nethird of the total requirements of the cashew factories.

There are three varieties if raw cashewnuts used in the factories at Mangalore. They are local nuts of Puttur, Vittal and other places which are uniformly large and good for out-turn and oil contents; African nuts, which are smaller than the local nuts and noted for the out-turn which are more by about 2 to 2.5 kgs but their oil content is less by about 10.0 per cent and Uttara Kannada nuts, which resemble those grown in Kundapur but are not so good as the local nuts. The cashewnut consists of a kernel covered with a thin pink skin enclosed in a double walled shell. The process of extracting kernels from raw cashewnuts are reasting, shelling, drying and peeling.

As the shell of the cashewnut is leathery, it is necessary for the rawnuts to be reasted before the kernels can be extracted from them. Until 1935, the roasting process resulted in the loss of cashew liquid which is a versatile products. It was only in 1935, that a process of roasting, whereby the cashew shell liquid could be extracted was evolved. The nuts are roasted at high temperature either in steel drums out of contact with air, or in cashewnut shell liquid oil leath (this process is known as hot oil bath process) to roast the outer hard shell and prepare them for being shelled. During the process of roasting, cashew kernel is extracted from the shell which, at the same time, is made brittle and easier to break. The roasted nuts coming out of the roasting plant have a small coating of cashew shell liquid which is removed by centrifuging before taking to the shellers for shelling.

After roasting, the nuts are to be shelled. This denotes the process of breaking open the outer thick shell and removing the kernel with its husk or peel attacked to its surface. This is done either by hand-shelling in which the roasted nuts are cracked open by beating with a wooden mallet or by machine shelling. The latter process is more economical and easy. This work requires patience and precision lest the kernels would be broken to pieces.

The shelled kernels have a pink skin covering which has to be removed before the kerncls are made edible. This skin is attached fast to the kernel and can only be loosened after the kernels have been dried and the moisture in the peel or husk is removed by contact with hot air. After the kernels are kept for several hours in a hot room or a drying chamber, they lose much of their moisture and shrink slightly. Thereafter the thin skin is removed by hand without difficulty. The process of removing the skins is known as 'peeling'. This is again a precision job done exclusively by female labour. The kernels, which are dried, are then peeled by hand by means of a blunt metallic knife. As in the case of shelling, peeling has to be done with one kernel at a time.

The peeled kernels are then graded into different grades depending on the size colour, appearance either wholes or pieces and such other considerations. Grading is adelicate operation, done manually by handpicking. The Export Inspection Agency has recommeded about 22 grades and all these grades are sold in the export market. The grades of cashew kernels very from 210 counts to 500 counts. Among the more important grades may be mentioned those of scorched wholes and desert wholes No. 1 and 2. The tips of the kernels of the grade of scorched whole are burnt a little. This is considered to be of a superior quality when compared to the desert wholes. Under desert wholes No. 1 there are two grades, namely, the Indian grade and the continental grade. These are said to be slightly inferior to those of scorched wholes. These are not exported to America. The grade called desert whole No. 2 is still inferior to the above grades and it is not also exported outside. Among the other grade may be mentioned the bits, splits, white large pieces, white small pieces, scorched pieces and desert pieces. From one bag of 50 kg of raw cashewnut about 11 to 12 kgs of the grades from 210 counts to 500 counts and the grade called scorched wholes, scorched wholes are obtained in addition to about 4.5 to 5.0 kgs of white pieces and desert pieces and about 1.5 to 2 kgs of scorched pieces. And from about of the same quantity, 8 kgs of cashew oil is extracted.

The cashew kernels are highly susceptible to infestation and insect attack when exposed to even ordinary atmospheric conditions in storage. The kernels are packed in sealed tins which are vaccumised and the vacuum is filled with carbondioxide. This process is known as the vitapack process. It ensures that infestation of the kernels inside is eliminated.

Cashew Shell Oil

Cashew shell liquid is an important by-product obtained while processing cashewnuts and extracting the kernels. During the process of roasting, most of the oil in the shell oozes out and this is collected. The shells obtained after shelling are crushed in expellers to separate the oil which is left in the shells. The commercial shell oil obtained by the roasting of cashewnuts in a dark brown viscous oil, easily soluble in most of the organic solvents. Although not a glyceride, it is drying oil with suitable driers, it gives a smooth, shining film of dark-brown solour. In India, the oil is used as a water-proofing agent and also as a preservation in the painting of boats, fishing nets and light wood work. It is only in recent years, mainly as a result of the investigations of M/S. Heavy and Caplan, that the oil has attained technical importance and America, Japan and U.K. are now the chief buyers of the oil produced in India. The industrial applications of shell oil are based upon its polymerization to a rubber-like material under the influence of adics, and on the formation of a wide range of condensation products with aldehydes. The latter are generally hard, infusible and extremely resistant to the action of chemicals such as acids and alkalies. The cashew shell liquid is now used in the manufacture of brake liners, paints, varnishes, lacquers and insulators.

The nut consists of the kernel (20-25%), the shell liquid (20-25%), the testa (two per cent) and the rest being the shell. Many refined extraction technique have of late been devised. Expeller method being used in Mangalore area is better than orher methods like hot oil bath method, kiln method, etc., as it extracts 90% of the oil. CNSL, extracted with low boiling petroleum, contains about 90 per cent anacardic acid and about 10 per cent cardol. CNSL, on distillation, gives the place yellow phenolic derivative cardanol. Natural CNSL contains 90.9 per cent anacardic acid, 10-15 cardol and small amounts of other materials notably the 2-methyl derivatives of cardol.

CNSL polymers are compatible with a wide range of polymers, such as plastics and rubbers. At present CNSL ploymers are being used in brake lining and clutch facing resins, because CNSL resins have the property of absorbing the heat generated by friction in the braking action while retaining their braking efficieney longer. CNSL based resins have outstanding resistance to the softening action of mineral oils and high resistance to acids and alkalies. Other notable properties include their resistance to "cold wear", good heat and electrical resistance, anti-microbial property, termite and insect resistance. Because of its dark colour CNSL is used in the manufacture of dard coloured paints and enamels. A number of anticorrossive paint formulations for ship bottoms have been made by the Regional Research Laboratory, Hyderabad, the Central Institute of Fisheries Technology, Cochin, Bombay University, and the Research, Design and Standards Organisation, Lucknow. Paints and varnishes made from CNSL have superior properties to those of conventional oil or synthetic resins. Varnishes resistant to water and gasoline have been made by incorporating sulphur in CNSL.

Electrical insulating varnishes were obtained by treating CNSL with formaldehyde and compounding the resulting material with pure phenolic resin varnish or alkyl resin in suitable proportions. Films of those materials are water and chemical resistant and can be used as insulating varnishes with high electrical resistance and as bobbin enamels and laboratory table tops.

Lacquers developed from CNSL could be used for insulation, protective or decorative coatings for furniture, buildings, automobiles, etc. Their films have toughness and elasticity, excellent gloss and superfine adhesive qualities. The dried films are superior to those of ordinary and paints in respect of resistance to oils, greases, moisture and chemicals. Cashew lacquers are cheaper than ordinary oil varnishes.

Cashew polymers react with formaldehyde of give a rubbery gel, which can be used as a cement hardening immune to acids and alkalies reaction. It can be used for cementing floors exposed to chemical attack.

CNSL modified by heating at 160-300°C in the presence of certain accelerators give stoving enamels resistant to alkali and acid solutions, mineral and fatty oils and various organic solvents. Coatings compositions possessing insecticidal properties were obtained by adding DDT, gammexane, etc, to CNSL or chlorinated CNSL after treatment with formaldehyde, gums and ressns and drying or semi-drying oils. CNSL or cardanol derivatives are extensively used in the laminating industry for reducing brittleness and improving the flexibility of the laminates. A CNSL-based adhesive for bending concrete to wooden surfaces was developed by the Central Building Research Institute, Roorkee. Adhesives suitable for plywood are made by oxidising CNSL with potassium permanganate or manganese dioxide at 100°C reacted with paraformaldehyde and compounded with $CUCT_2$. Also CNSL modified with furfural, aniline, xylol, etc. gives good plywood adhesives. CNSL is also used as substitute for linseed oil in the manufacture of foundry core oil, which is used as a binder in the foundry.

Cardanol and its derivates can also be converted to phenoplasts with better processability, hydrocarbon solubility and resistance to acids and alkalies than the conventional phenol-based systems. Moulding powers from CNSL, shallac, and fillers, such as wood flour, sawdust, asbestos, etc. are found to give article with excellent finish, good flexural and tensile strengths and satisfactory water resistance. Stable rigid or flexible covering materials in the from of titles sheets, etc. were made from compositions containing CNSL, formalin, natural rubber or synthetic rubber and other conventional ingredients. Light weight, sandwich type plastics, composite panels suitable for partitions, claddings, flush doors, etc. have been developed using resins based on CNSL. Foam plastics based on CNSL and its derivatives have also been made.

The use of CNSL in rubber compositions has been found to improve the performance of rubber products. It helps processing and enhances the vulcanizate properties. CNSL enhances the insolubility of natural rubber vulcanizates in petroleum solvents. It helps in the incorporation of ingredients into rubber and increases it resistance to moisture.

CNSL Trends

The cashew shell (peri-carp) contains humidity (13.17%), ashes (6.74%), cellulose lignic (17 35%), azotic substances (1.06%), saccharificable substances (20.85%) and substances soluble in ether (35.10%). Since by hot oil both method, only about 7.12 per cent of the nut weight is recoverable as CNSL, improvements have to be done to recover the maximum of 20-25 per cent by solvent extraction or expulsions.

Kerala and Karnataka are major CNSL producers. Export of CNSL has been about 6.3 per cent of total nut production during 1962 ; it declined to 5.5 per cent during 1970-72 and increased 7.0 per cent in 1975. (Table 15.1) But in recent times, the export has fallen considerably. Therefore, internal market sources and diversification of its uses have to be attempted. Substitutes like mineral phenols have reduced the demand for CNSL for disc brakes in the U.S.A. So, to keep the export trends of CNSL at par with that of cashew kernels, it becomes necessary to diversify its markets and uses Table 15.2 presents consumption of CNSL by various industries. Table 15.3 describes the quantity of CNSL exported from Mangalore port which is declining very much in recent years.

The cashew kernels are packed in tins of 11.34 kgs, though cost of production is usually calculated for bags of 74 kgs of rawnuts. The proportion of exports ranges from nil in some factories in recent years (both kernels and CNSL) to over 50 per cent of kernels in others. Most of the factories do not export CNSL of late, but a few factories report that during 1986-87, pending orders could be exported to the U.S.A and other countries. The capital investment ranges from Rs. 10.0 to Rs. 20.0 lakhs. Of the total labour employed, temporary labour, accounts for a substantial proportion (about 40 to 45 per cent). Female labour accounts for over 90.0 per cent of total labour indicating the crucial role played by women in cashew processing. The wage rate ranges from Rs. 12.00 to Rs. 20.00 for different types of labour. Some factories report an equal quantity of cashewnut kernel production and CNSL export.

Table 15.4 presents the cost of production of cashew kernels and cashewnut shell liquid for a typical factory. The cost of production per bag of 75 kgs of rawnut comes to Rs. 1280. Of this, the cost of rawnut is Rs. 14.40 per kg, cost of processing Rs. 100 per bag and interest, selling expenses, etc. Rs. 100.

From the processed rawnuts, 3 grades of kernels are obtained—wholes 20.0 per cent, pieces 4.0 per cent and shells 73.3 per cent. Wholes fetch Rs. 75.0 per kg. Thus, net profit per bag comes to Rs. 80.0.

From the same Table, cost of production and yield of cashewnut shell liquid are also described. The cost and yield structure is for one metric tonne. The cost of production is Rs. 1150 and yield Rs. 1390 at 25 per cent of recovery. The net profit works out to Rs. 240 per mt of cashew shell processing.

In recent times, extraction of CNSL has been coming down owing to international competition, lowering demand, decreasing prices, cheap substitutes, etc. So, it has become paramount to seek alternative markets, reduce cost of production, increase internal consumption and diversify the uses of CNSL so that its extraction becomes profitable. Only when the processing of casehwnut shell liquid is combined with kernels that the cashew processing industry would becomes economically viable.

TABLE 15.1 : **Production of CNSL in India**

(Quantity in metric tonnes)

Year	*Kerala*	*Karnataka*	*Tamil Nadu*	*Total*
1970	4,600	4,200	1,000	9,800
1971	4,200	3,800	900	8,900
1972	3,500	3,800	800	8,100
1973	3,500	3,900	700	8,100
1974	9,500	3,900	600	14,000

Source: 'Cashewnut shell liquid—Prospects for Expansion of Internal Consumption and Exports' by Indian Institute of Foreign Trade, p. 13.

TABLE 15.2 : **Consumption of CNSL by various industries**

(Quantity in metric tonnes)

Industrial uses	*1974*	*1975*	*1976*	*1977*	*1978*
Brake linings	1000	1100	1200	1350	1500
Paints and varnishes	900	1100	1300	1600	1900
Chemical resistant cement	60	75	95	120	130
Oil tempered hardboards	60	50	70	80	100
Foundry core oil	700	850	1050	1300	1600
Epoxy resins	neg—included in the estimate for cardonal				
Laminates			-do-		
Water proofing compound	16	20	20	30	32
Filter paper	—	150	150	150	150
CNSL based resins	20	50	50	50	50
Cardanol	850	1100	1340	1900	2280
Total	3596	4495	5275	6580	7762

Source: 'CNSL Prospects for Expansion of Internal Consumption and Exports' by the Indian Institute of Foreign Trade, p. 45.

TABLE 15.3 : Quantity of CNSL exported from Mangalore

Year	*Qty. exported (mt)*	*Rs./tonne*
1979-80	356.0	8096.7
1980-81	300.00	4680.0
1981-82	1625.0	3619.0
1982-83	1790.0	3744.5
1983-84	750.0	6040.4
1984-85	457.0	—

Source : Kanara Chamber of Commerce and Industry, Mangalore.

TABLE 15.4 : Cost of Kernels and CNSL

	Rs. P
A. Cost of Production per bag of 75. kg	
Cost of rawnuts per bag of 75 kgs	1080
Cost of processing	100
Interest on working capital, term loan and other borrowings	90
Selling and ather expenses	10
Sub-total	1280
B. Net Output	
Wholes 15 kgs @ Rs. 75 per kg	1125
Pieces 3 kgs @ Rs. 60 per kg	180
Shells 55 kgs @ Re. 1 per kg	55
Sub-total	1360
Net perofit per bag	80
C. Cost of production of CNSL per mt.	
Cost of 1 mt. shells @ Re. 1 per kg	1000
Production cost	150
Sub-total	1150
Yield : 25% i.e. 250 kgs @ Rs. 4 per kg	1000
Shell cake 650 kgs @ Re. 0.6/kg	390
Sub-total	1390
Net profit per mt.	240

Source : Oriental Cashew Industries, Puttur.

16

Cashew Apple

Introduction

Cashew apples (*Anacardium occidentale*) are an important by-product of cashewnut industry. The cashew plantations are mostly concentrated in the states of Kerala, Tamil Nadu, Maharashtra, Karnataka, Andhra Pradesh, Goa and Orissa. Total production of cashewnut and cashew apples in India during 1975-76 was reported to be 1.7 and 8.4 lakh tonnes and current production is estimated to be 1.9 and 9.5 lakh tonnes, respectively. For every tonne of nut produced roughly except for a small quantity (about 15%) used for direct consumption or for production of country liquor. This means roughly 8.0 lakh tonnes of fruits are wasted. By taking a conservative price of 50 paise per kg of fruit, the annual loss to the horticultural wealth of the country is estimated to be Rs. 400 million. There is, therefore, the necessity to find ways and means of utilizing the fruit to produce various products and thus create commercial outlet for these fruits. This will also add to the profitability of cashewnut industry.

The fruits are highly susceptible to injury and rapid microbial spoilage. Recently, during the field trials as high as 65% of the fruits were found to spoil within 24 hours of harvesting and transportation from the field to the processing centre. The fruits are, therefore, required to be pretreated to prevent their rapid spoilage.

Cashew apples contain astringent and acrid principles which produce an unpleasant biting sensation on the tongue and throat when eaten as such. These factors together with the difficulties experienced in collection of fruits have come in the ways of commercial utilisation of this nutritious fruit.

However, as a result of investigations carried out, it has been found possible to remove astringent and acrid principles by simple methods and utilise these fruits for conversion into several products. These products include cashew apple juice, juice beverages, blended juice beverages, cashew apple juice concentrates, cashew-apple preserve and candy, cashew apple vinegar, cashew apple jam or mixed fruit jam, cashew apple pickles.

Cashew-apple is not a true fruit but a swollen peduncle to which cashewnut is attached. It is a soft but fibruous juicy fruit. It possesses exotic flavour characteristic of the fruit. Based on external colour of the fruit cashew appled can be broadly classified into two varieties namely Red and Yellow.

The fruit weight varies from 26 to 50 gms and the yield of juice 60-70%. The pH of the juice ranges from 3.9 to 4.1. The total soluble solids mostly consisting of reducing sugars range from 12-13% and the acidity from 0.2% to 0.35%. The astringent and acrid taste of the fruit has been attributed to its tannins and an oily substance is not identified. Leucodelphinidin has been identified as the major polyphenol constituent of cashew apple juice. Ascorbic acid and total polyphenol content are the two important factors from the point of view of processing cashew appled. The tannin content varies from 0.1 to 0.7%. The ascorbic acid content varies from 120 mg to 306 mg.

As already pointed out, cashew appled are highly perishable commodities. Dipping the fruits in preservative solutions (1% KMS+1% sodium benzoate+0.1% citric acid) for 5 minutes and packing the fruits in bamboo baskets provided with sufficient windows and cushioning material at the bottom was found to be very effective in reducing the spoilage of fruits to practical limits during transportation and storage for 24 hours. Alternately steeping the fruits in preservative solutions (0.2%KMS+0.05% citric acid) was found to be effective in preventing the spoilage completely. But the added weight is a drawback during transportation. The former method of dip treatment was, therefore, considered to be practical.

Various methods have been tried to remove those disagreeable principles. Out of these, three methods have been found to be satisfactory in the preparation of different products. These include (i) steaming the fruits for 5 minutes under the steam pressure of 5 psig and subsequent washing of the fruits with cold water, (ii) cooking the fruits for 5 minutes in boiling solution of common salt (2%) and (iii) by addition of required quantity of gelating solution to the expressed juice.

Good and sound fruits should be used for juice extraction. The juice can be extracted from cut fruits in a screw type juice extractor or by pressing in basket press preferably by a combination of two operations to obtain maximum yield. The stringent and acrid principles are removed by gelatin treatment (concentration of gelatin in treated juice to be 0.5%). After settling and filteration the brix of the juice is raised to 15 and acidity to 0.4% by addition of sugar and citric acid. The juice thus prepared is then bottled. A carbonated cashew apple juice beverage of highly acceptability can also be prepared keeping the juice content at 50%, brix at 15, acidity at 0.3% and carbonating the drink to 2.5 volume level.

Some consumers do not very much like the cashew apple flavour. So attempts have been made to blend the cashew-apple juice with other fruit juices, like grape, apple, mango, pineapple, papaya, orange, lime etc. Cashew apple juice was found to blend better only with pineapple and lime juices. Other blends although found to give good product to begin with, during storage developed off-flavour. Beverages prepared based on cashew apple and pineapple juice blend were found to be the most acceptable dring in different regions of India.

A good quality cashew appled juice concentrate could be prepared by concentrating sulphited clarified juice in a forced circulation evaporator. It has been found to be not advantageous to concentrate the juice beyond 60 brix.

Processing of the fruits

The fruits are collected from the orchard after separating the nuts, and transported to the processing unit. As the fruits get damaged during plucking, transportation is to be done as quickly as possible to avoid growth of undesirable micro-organisms. The fruits are washed in running tap water and processed through fruit mill. The pulp obtained from the fruit mill is mixed with potassium metabisulphite to obtain 100 to 150 ppm of SO_3. The juice is extracted by passing the pulp through screw press. Yield of juice varies from 55 to 60 per cent.

After sufficient maturation the strength of the distillate is adjusted to 43% by addition to water. Little amount of caramel is added to get the colour (only caramel is permitted). It is filtered, bottled and labelled. The bottle and label should be attractive to have consumer eye appeal. The chemical composition of fenni is given below.

Composition of Fenni

Constituents	*Composition*
Ethyl Alcohol	43.00%
Total solids	0.138%
Volatile acidity	19.0 gm/100 L. of A.A.*
Esters	244.2 gm/100 L. of A.A
Aldehydes	28.0 gm/100 L. of A.A
Higher alcohol	196.2 gm/100 L. of A.A
Furfural	5.0 gm/100 L. of A.A

*Expressed as grams/100 litres of absolute alcohol.

Cashew apple fruits after collection are crushed and the juice is expressed by either a screw press or basket press. The juice is then sulphited to destroy the bacteria. Cashew apple residue left after juice extraction accounts for 30-40 per cent of the whole fruit. The residue contains on dry weight basis about 9 per cent protein, 4 per cent fat, 8 per cent crude fibre, 10 per cent pectin and also rich in ascorbic acid minerals. The residue has been suggested to be utilised for the recovery of low methoxyl pectin (LMP) as its gelly grade has been found to be poor or as cattle feed after drying.

Adding cashew kernel meal, kola chocolate, plam cocoa chocolate and a chocolate in the proportion 1 : 1 : 1 of cashew-kernel meal, kola powder and cocoa could be manufactured. These three products are known as 'kocho', 'cocho' and 'camcho' respectively. Consumer trials have shown that all the three products are preferred by the consumers. Of the three 'camcho' has the highest rating. Therefore, the recently established chocolate factory at Puttur may consider using cashew kernel to prepare 'camcho' chocolate.

Cashew apple contains high protein value and could even compete with mango and other citrus fruits in containing ascorbic acid, thiamin and other B vitamins. Pectin from cashew apple juice can be used in ice-cream, jellies sweets and jam. From the juices if tannin is removed, it will have wide usage.

If we compare the project economics for cashew apple wine (Rs. 11.00 per bottle of 750 cc) and cashew apple brandy (Rs. 25.75 per bottle of 750 cc), it works out to be quite cheap (Table 16.1). So it could be better if through cooperative enterprise, a cashew apple processing factory is established in this district.

For raw materials, the factory can look to other districts and Kerala. It becomes paramount that the farmers are encouraged to pluck only ripe fruits and market them through the industrial cooperative which may be started to look into all interests of cashew industry. When the Goan cashew cultivators obtain nearly 35 per cent of their income through cashew apples, a similar situation here is highly recommended.

Table 16.1 : Project Economics for Cashew apple products

Particulars		*Cashew apple Wine*	*Cashew apple Brandy*
I. Capacity (tonnes/shift)		4.0	4.4
Working days (p.a)		70	100
Output (750 cc bot/p.a)		60,000	23,000
II. Capital investment:			
(i) Land & buildings	Rs.	3.56 lakhs	3.16 lakhs
(ii) Plant & machinery	Rs.	8.69 ,,	6.47 ,,
(iii) Working capital	Rs.	4.11 ,,	2.66 ,,
Total	Rs.	16.36 ,,	12.29 ,,
III. Cost of production:			
(i) Raw materials, etc.	Rs.	2.90 ,,	1.84 ,,
(ii) Utilities	Rs.	0.33 ,,	0.20 ,,
(iii) Labour & Supervision	Rs.	3.11 ,,	3.38 ,,
Total	Rs.	7.20 ,,	6.60 ,,
Cost per bottle of 750 cc	Rs.	12.00 ,,	28.70 ,,

Source : N.V. Sripathy, 'Cashew Apple Products' in Proceedings of the Seminar on Cashew Research and Development in the Coastal Region, ARS (UAS), Ullal, 1985.

17

Cashew : An Overview

Cashewnut is an important foreign exchange earning crop. During the mid-seventies, the growth rate of area under the crop declined but due to centrally sponsored and state schemes, the area has begun to spurt. The World Bank aided Multi-State Project has been responsible for area increase, widespread introduction of improved varieties and methods of cultivation and increase in production. The major problems cashew cultivation and the industry face are declining yield rates, falling export performance (due to declining rawnut imports also), rising cost of cultivation and lack of full-scale adoption of improved methods. Not only there is a fall in cashew kernel exports (through India is still to leading exporter) due to international competition, there has been a decline in the exports of cashewnut shell liquid also. Therefore internal market has to be intensified for the cashewnut shell liquid. Cashew apple has been a neglected sector of cashew industry and efforts have to be done toward better utilisation of the apple into various utility products. Processing of cashew apple could be integrated with cashewnut processing.

The evaluation study on the Multi-State Cashew Project has been undertaken in the Dakshina Kannada district covering 150 subsidy farmers and 100 non-subsidy farmers in Bantwal, Belthangady, Karkal, Mangalore, Puttur and Udupi taluks during 1985-86 The average cashew area has been about 26 5 and 23.0 per cent of total cropped area in the subsidy and non-subsidy samples respectively. Tree population of improved cashew varieties (Ullal I and Ullal II) introduced during the World Bank Project period (1980-81 to 1986-87) has been 200 trees per ha, though there are a slight variations among the individual adopters. Some of the non-subsidy farmers also adopted the improved varieties. The tree population was intensive (75 to 86 per cent) for the improved varieties in the 3-5 years ago, indicating early adoption by the farmers.

Whereas in local varieties, the above 5-10 years groups dominate the scene. Above 20 year old trees share a substantial proportion in Bantwal, Belthangady and Mangalore. In the non-subsidy cases, the proportion under older trees is higher. The proportion of yielding trees is very low for the less than 5 years age group for both local and improved varieties (usually less than 10.0 per cent). Even in the above 5 year group, the local varieties report only upto 80.0 capacity. The means of the unproductive trees from a substantial proportion.

Among the subsidy farmers the average yield for less that 3 year old trees is 46.0 kgs/ha for local and 60 kgs/ha for the improved varieties. The yield rates increase to 330 kgs/ha and 480 kgs/ha in 5-10 years group and 740.0 kgs/ha and 800/ha for the 5-10 year group. The yield rate of improved varieties has been higher than that of the local varieties. For the non-subsidy case, the yield rates have been 44.0 kgs/ha and 726 kgs/ha respectively, indicating a slightly lower rate. But for old trees, the yield rate in the non-subsidy cases is slightly higher.

Manures and chemical fertilisers are being applied more for improved than local varieties. For improved varieties, the per tree application of manure has been 1.63 kgs and fertilisers 0.68 kgs respectively. For locals, the application has been 1.5 kgs and 0.17 kgs per tree. Plant protection measures are also being adopted, but fertiliser and insecticide application has yet to be intensified.

The productivity trends for other crops grown along with cashewnut show that, the subsidy cases register a slight increase in paddy and coconut yield, whereas the non-subsidy units have higher yield rates in arecanut.

It has been found that on average 15.0 per cent of tender nuts are plucked in subsidy and 13.2 per cent in non-subsidy samples, indicating the seriousness of the problem in future. Cultural practices alike pruning is being adopted, though top working is almost not ried or unknown among the farmers.

Altogether, three grades of rawnut are marketed through private mechanism and the proportion of first quality is less than 25.0 per cent in Karkal, Puttur and Udupi for subsidy cases and in Belthangady, Karkal and Mangalore for non-subsidy cases. The medium and poor qualities find a high share in almost all places. The loan per acre received by the farmers been Rs. 1578.0 in Bantwal and Rs. 1421.0 in Mangalore. The subsidy is uniform at Rs. 360.0 per acre in all the places.

In D.K. the cashewnut area has increased from 18 692 to 29,091 ha during the Project period. About 10,000 ha of waste-land has been brought under improved cashew cultivation. The production of cashewnut has increased from 13,085 mt to 14,277, indicating a declining yield rate. By 1990-91 when the improved trees would be in a mature fruit-bearing stage. The per ha productivity would be increasing substantially.

Of the taluks, Karkal has shown the largest additional area, through in Udupi and Puttur where large area expansion was envisaged, the increase has not been substantial. During the project period about 8000 beneficiaries covering about 33.3 per cent of total cashew area has been benefited. The primary land development banks have been the major source of finance.

Suggestions

Since the yield rate in Karnataka is low (when compared to say, Kerala) cashew research and application has to be strengthened toward effective application of chemical fertilisers, application of plant protection chemicals, improvement of old trees by top working and pruning and plantation hygiene. For this, extension work has to be spreaded up and adoption of farmers for chemical demonstrations, top working, etc., have to be conducted regularly.

Commercial banks, cooperatives and voluntary institutions have to come forward to assist the farmers financially and technically. The World Bank Project may be extended under State and Central assistance programmes. Since only about one-third of cashew farmers and area has been covered so far, inclusion of others and intensified assistance to the already covered assume greater importance. They have to be properly educated in rejuvenating the old trees and adoption of cultural practices toward commercialising cashew cultivation. Plucking of tendernuts has to be thoroughly discouraged.

The processing factories, in order to realise full capacity utilisation have to come forward to encourage the farmers toward better production. May could attempt to integrate processing of cashew shell oil and cashew apples along with cashewnut to realise full exployment and economics of scale.

Cooperative marketing among the growers may be encouraged for effective marketing and for better storage and disposal. Incentives have to be given toward this direction. The harvesting period may be extended.

Price fluctuations and tax structure have to be rationalised to allow little differences among the cashew growing states.

A survey of unproductive trees has to be undertaken and rejuvenation of the same has to be attempted. Illegal felling of productive trees has to be discouraged.

Better cost economics has to be introduced to reduce cost of production which would enable effective competition against other countries in export performance. Remunerative and stable price rulings have to be effected toward this direction.

For effective utilisation of wastelands, better coordination between the forest, revenue and horticulture departments and establishing a Wasteland Utilisation Authority is essential;

To improve the tempo of fertiliser use and plant protection measures among the beneficiaries, fertiliser demonstrations, training and extension are necessary; For this a separate extension wing could be established which could also give guidance in top working, ripe fruit plucking and improving the quality;

For the non-beneficiaries also, extension is required and an increasing number of them are to be covered under the subsidy programme;

A larger participation of commercial banks in assisting the cashew cultivators would be a welcome feature and special cells may be created in the commercial banks to look after the interests of cashew cultivators and industrialists;

Since manufacture of cashew kernel, cashew shell oil and

utilisation of cashew apple are integrative, attempts have to be made to encourage better marketing for cashew shell oil and effective processing of cashew apple. Small scale manufacturers are to be assisted in realising the above, especially with regard to utilisation of cashew apple.

Finally, the project's vigour could be continued even after its completion by suitably providing the basic infrastructural facilities and incentives by the horticulture department. In this, research and development into improved varieties and methods of cultivation, training and extension, improvement of quality, better utilisation of cashew by-products, etc, are called for.

Index